John Brockman is the founder and publisher of the influential science salon Edge.org. He is the CEO of the literary agency Brockman Inc., and lives in New York City.

www.transworldbooks.co.uk

Also by John Brockman

As Author:
By the Late John Brockman
37
Afterwords
The Third Culture: Beyond the Scientific Revolution
Digerati

As Editor:
About Bateson
Speculations
Doing Science
Ways of Knowing
Creativity
The Greatest Inventions of the Past 2,000 Years
The Next Fifty Years
The New Humanists
Curious Minds
What We Believe but Cannot Prove
My Einstein
Intelligent Thought
What Is Your Dangerous Idea?
What Are You Optimistic About?
What Have You Changed Your Mind About?
This Will Change Everything
Is the Internet Changing the Way You Think?

As Co-editor:
How Things Are (*with Katinka Matson*)

This Will Make You Smarter

Edited by

JOHN BROCKMAN

Foreword by David Brooks

BLACK SWAN

TRANSWORLD PUBLISHERS
61–63 Uxbridge Road, London W5 5SA
A Random House Group Company
www.transworldbooks.co.uk

THIS WILL MAKE YOU SMARTER
A BLACK SWAN BOOK: 9780552778480

First published in Great Britain
in 2012 by Doubleday
an imprint of Transworld Publishers
Black Swan edition published 2012

A CIP catalogue record for this book
is available from the British Library.

Addresses for Random House Group Ltd companies outside the UK
can be found at: www.randomhouse.co.uk
The Random House Group Ltd Reg. No. 954009

The Random House Group Limited supports The Forest Stewardship Council (FSC®),
the leading international forest-certification organization. Our books carrying the
FSC label are printed on FSC®-certified paper. FSC is the only forest-certification scheme
endorsed by the leading environmental organizations, including Greenpeace.
Our paper-procurement policy can be found at www.randomhouse.co.uk/environment

Typeset in Janson
Printed and bound by CPI Group (UK) Ltd, Croydon, CR0 4YY.

2 4 6 8 10 9 7 5 3 1

CONTENTS

Foreword

DAVID BROOKS
Columnist, New York Times; *author*, The Social Animal

Every era has its intellectual hotspots. We think of the Bloomsbury Group in London during the early twentieth century. We think of the New York intellectuals who wrote for little magazines like *Partisan Review* in the 1950s. The most influential thinkers in our own era live at the nexus of the cognitive sciences, evolutionary psychology, and information technology. This constellation of thinkers, influenced by people like Daniel Kahneman, Noam Chomsky, E. O. Wilson, Steven Pinker, Steve Jobs, and Sergey Brin, do a great deal to set the intellectual temper of the times. They ask the fundamental questions and shape debates outside of their own disciplines and across the public sphere.

Many of the leaders of this network are in this book. They are lucky enough to be at the head of fast-advancing fields. But they are also lucky enough to have one another. The literary agent and all-purpose intellectual impresario John Brockman gathers members of this network for summits. He arranges symposia and encourages online conversations. Through Edge.org, he has multiplied the talents of everybody involved. Crucially, he has taken scholars out of their intellectual disciplines, encouraging them to interact with people in different fields, to talk with business executives, to talk with the general public.

The disciplinary structure in the universities is an important foundation. It enforces methodological rigor. But it doesn't really correlate with reality (why do we have one field, psychology, concerning the inner life and another field, sociology, concerning the outer life, when the distinction between the two is porous and maybe insignificant?). If there's going to be a vibrant intellectual life, somebody has to drag researchers out of their ghettos, and Brockman has done that, through *Edge*.

The book you hold in your hand accomplishes two things, one implicit, one explicit. Implicitly it gives you an excellent glimpse of what some of the world's leading thinkers are obsessed with at the moment. You can see their optimism (or anxiety) about how technology is changing culture and interaction. You'll observe a frequent desire to move beyond deductive reasoning and come up with more rigorous modes of holistic or emergent thinking.

You'll also get a sense of the emotional temper of the group. People in this culture love neat puzzles and cool questions. Benoit Mandelbrot asked his famous question "How long is the coast of Britain?" long before this symposium was written, but it perfectly captures the sort of puzzle people in this crowd love. The question seems simple. Just look it up in the encyclopedia. But as Mandelbrot observed, the length of the coast of Britain depends on what you use to measure it. If you draw lines on a map to approximate the coastline, you get one length, but if you try to measure the real bumps in every inlet and bay, the curves of each pebble and grain of sand, you get a much different length.

That question is intellectually complexifying but also clarifying. It gets beneath the way we see, and over the past generation the people in this book have taken us beneath our own conscious thinking and shown us the deeper patterns and realms of life. I think they've been influenced by the ethos of Silicon Valley. They seem to love heroic attempts at innovation and don't believe there is much disgrace in an adventurous failure. They are enthusiastic. Most important, they are not coldly deterministic. Under their influence, the cognitive and other sciences have learned from novels and the humanities. In this book, Joshua Greene has a brilliant entry in which he tries to define the relationship between the sciences and the humanities, between brain imaging and *Macbeth*. He shows that they are complementary and interconnected magisteria. In this way the rift between the two cultures is being partially healed.

The explicit purpose of this book is to give us better tools to think about the world. Though written by researchers, it is eminently practical for life day to day.

As you march through or dance around in this book, you'll see that some of the entries describe the patterns of the world. Nicholas Christakis is one of several scholars to emphasize that many things in the world have properties not present in their parts. They cannot be understood simply by taking them apart; you have to observe the interactions of the whole. Stephon Alexander is one of two writers (appropriately) to emphasize the dualities found in the world. Just as an electron has both wave-like and particle-like properties, so many things can have two sets of characteristics simultaneously. Clay Shirky emphasizes that while we often imagine bell curves everywhere, in fact the phenomena of the world are often best described by the Pareto Principle. Things are often skewed radically toward the top of any distribution. Twenty percent of the employees in any company do most of the work, and the top 20 percent within that 20 percent do most of that group's work.

As you read through the entries that seek to understand patterns in the world, you'll run across a few amazing facts. For example, I didn't know that twice as many people in India have access to cell phones as to latrines.

But most of the essays in the book are about metacognition. They consist of thinking about how we think. I was struck by Daniel Kahneman's essay on the Focusing Illusion, by Paul Saffo's essay on the Time Span Illusion, by John McWhorter's essay on Path Dependence, and Evgeny Morozov's essay on the Einstellung Effect, among many others. If you lead an organization, or have the sort of job that demands that you think about the world, these tools are like magic hammers. They will help you, now and through life, to see the world better, and to see your own biases more accurately.

But I do want to emphasize one final thing. These researchers are giving us tools for thinking. It sounds utilitarian and it is. But tucked in the nooks and crannies of this book there are insights about the intimate world, about the realms of emotion and spirit. There are insights about what sort of creatures we are. Some of these are not all that uplifting. Gloria Origgi writes about Kakonomics, our preference for low-quality outcomes. But Roger Highfield, Jonathan Haidt, and others write about the "snuggle for existence": the fact that evolution is not only about competition, but profoundly about cooperation and even altruism. Haidt says wittily that we are the giraffes of altruism. There is something for the poetic side of your nature, as well as the prosaic.

The people in this book lead some of the hottest fields; in these pages they are just giving you little wisps of what they are working on. But I hope you'll be struck not only by how freewheeling they are willing to be, but also by the undertone of modesty. Several of the essays in this book emphasize that we see the world in deeply imperfect ways, and that our knowledge is partial. They have respect for the scientific method and the group enterprise precisely because the stock of our own individual reason is small. Amid all the charms to follow, that mixture of humility and daring is the most unusual and important.

PREFACE: THE *EDGE* QUESTION

JOHN BROCKMAN
Publisher and editor, Edge

In 1981 I founded the Reality Club. Through 1996, the club held its meetings in Chinese restaurants, artists' lofts, the boardrooms of investment-banking firms, ballrooms, museums, and living rooms, among other venues. The Reality Club differed from the Algonquin Round Table, the Apostles, or the Bloomsbury Group, but it offered the same quality of intellectual adventure. Perhaps the closest resemblance was to the late eighteenth- and early nineteenth-century Lunar Society of Birmingham, an informal gathering of the leading cultural figures of the new industrial age—James Watt, Erasmus Darwin, Josiah Wedgwood, Joseph Priestley, Benjamin Franklin. In a similar fashion, the Reality Club was an attempt to gather together those people exploring the themes of the postindustrial age.

In 1997, the Reality Club went online, rebranded as *Edge*. The ideas presented on *Edge* are speculative; they represent the frontiers in such areas as evolutionary biology, genetics, computer science, neurophysiology, psychology, and physics. Emerging out of these contributions is a new natural philosophy, new ways of understanding physical systems, new ways of thinking that call into question many of our basic assumptions.

For each of the anniversary editions of *Edge*, I have asked contributors for their responses to a question that comes to me, or to one of my correspondents, in the middle of the night. It's not easy coming up with a question. As the late James Lee Byars, my friend and sometime collaborator, used to say: "I can answer the question, but am I bright enough to ask it?" I'm looking for questions that inspire answers we can't possibly predict. My goal is to provoke people into thinking thoughts they normally might not have.

This year's question, suggested by Steven Pinker and seconded by Daniel Kahneman, takes off from a notion of James Flynn, intelligence researcher and emeritus professor of political studies at the University of Otago in Dunedin, New Zealand, who defined *shorthand abstractions* (SHAs) as concepts drawn from science that have become part of the language and make people smarter by providing widely applicable templates. "Market," "placebo," "random sample," and "naturalistic fallacy" are a few of his examples. His idea is that the abstraction is available as a single cognitive chunk, which can be used as an element in thinking and in debate.

The *Edge* Question 2011

What Scientific Concept Would Improve Everybody's Cognitive Toolkit?

Here, the term "scientific" is to be understood in a broad sense—as the most reliable way of gaining knowledge about anything, whether it be human behavior, corporate behavior, the fate of the planet, or the future of the universe. A "scientific concept" may come from philosophy, logic, economics, jurisprudence, or any other analytic enterprises, as long as it is a rigorous tool that can be summed up succinctly but has broad application to understanding the world.

This Will Make You Smarter

"DEEP TIME" AND THE FAR FUTURE

MARTIN REES

President emeritus, the Royal Society; professor of cosmology & astrophysics; master, Trinity College, University of Cambridge; author, Our Final Century: The 50/50 Threat to Humanity's Survival

We need to extend our time horizons. Especially, we need deeper and wider awareness that far more time lies ahead than has elapsed up until now.

Our present biosphere is the outcome of about 4 billion years of evolution, and we can trace cosmic history right back to a Big Bang that happened about 13.7 billion years ago. The stupendous time spans of the evolutionary past are now part of common culture and understanding—even though the concept may not yet have percolated to all parts of Kansas and Alaska. But the immense time horizons that stretch ahead—though familiar to every astronomer—haven't permeated our culture to the same extent.

Our sun is less than halfway through its life. It formed 4.5 billion years ago, but it's got 6 billion more years before the fuel runs out. It will then flare up, engulfing the inner planets and vaporizing any life that might then remain on Earth. But even after the sun's demise, the expanding universe will continue, perhaps forever—destined to become ever colder, ever emptier. That, at least, is the best long-range forecast that cosmologists can offer, though few would lay firm odds on what may happen beyond a few tens of billions of years.

Awareness of the "deep time" lying ahead is still not pervasive. Indeed, most people—and not only those for whom this view is enshrined in religious beliefs—envisage humans as in some sense the culmination of evolution. But no astronomer could believe this; on the contrary, it would be equally plausible to surmise that we are not even at the halfway stage. There is abundant time for posthuman evolution, here on Earth or far beyond, organic or inorganic, to give rise to far more diversity and even greater qualitative changes than those that have led from single-celled organisms to humans. Indeed, this conclusion is strengthened when we realize that future evolution will proceed not on the million-year time scale characteristic of Darwinian selection but at the much accelerated rate allowed by genetic modification and the advance of machine intelligence (and forced by the drastic environmental pressures that would confront any humans who were to construct habitats beyond the Earth).

Darwin himself realized that "not one living species will preserve its unaltered likeness to a distant futurity." We now know that "futurity" extends far further—and alterations can occur far faster—than Darwin envisioned. And we know that the cosmos, through which life could spread, is far more extensive and varied than he envisioned. So humans are surely not the terminal branch of an evolutionary tree but a species that emerged early in cosmic history, with special promise for diverse evolution. But this is not to diminish their status. We humans are entitled to feel uniquely important, as the first known species with the power to mold its evolutionary legacy.

WE ARE UNIQUE

MARCELO GLEISER
Appleton Professor of Natural Philosophy and professor of physics and astronomy, Dartmouth College; author, A Tear at the Edge of Creation: A Radical New Vision for Life in an Imperfect Universe

To improve everybody's cognitive toolkit, the required scientific concept has to be applicable to all humans. It needs to make a difference to us as a species, or, more to the point I am going to make, as a key factor in defining our collective role. This concept must affect the way we perceive who we are and why we are here. It should redefine the way we live our lives and plan for our collective future. This concept must make it clear that we matter.

A concept that might grow into this life-redefining powerhouse is the notion that we, humans on a rare planet, are unique and uniquely important. But what of Copernicanism, the notion that the more we learn about the universe the less important we become? I will argue that modern science, traditionally considered guilty of reducing our existence to a pointless accident in an indifferent universe, is actually saying the opposite. Whereas it does say that we are an accident in an indifferent universe, it also says that we are a rare accident and thus not pointless.

But wait! Isn't it the opposite? Shouldn't we expect life to be common in the cosmos and us to be just one of many creatures out there? After all, as we discover more and more worlds circling other suns, the so-called exoplanets, we find an amazing array of possibilities. Also, given that the laws of physics and chemistry are the same across the universe, we should expect life to be ubiquitous: If it happened here, it must have happened in many other places. So why am I claiming that we are unique?

There is an *enormous* difference between life and intelligent life. By intelligent life, I don't mean clever crows or dolphins but minds capable of self-awareness and of developing advanced technologies—that is, not just using what's at hand but transforming materials into devices that can perform a multitude of tasks. I agree that single-celled life, although dependent on a multitude of physical and biochemical factors, shouldn't be an exclusive property of our planet—first, because life on Earth appeared almost as quickly as it could, no more than a few hundred million years after things quieted down enough; and second, because the existence of extremophiles, life-forms capable of surviving in extreme conditions (very hot or cold, very acidic or/and radioactive, no oxygen, etc.), show that life is resilient and spreads into every niche it can.

However, the existence of single-celled organisms doesn't necessarily lead to that of multicellular ones, much less to that of *intelligent* multicellular ones. Life is in the business of surviving the best way it can in a given environment. If the environment changes, those creatures that can survive under the new conditions will. Nothing in this dynamic supports the notion that once there's life all you have to do is wait long enough and *poof!* up pops a clever creature. This smells of biological teleology, the concept that life's purpose is to create intelligent life, a notion that seduces many people for obvious reasons: It makes us the special outcome of some grand plan. The history of life on Earth doesn't support this evolution toward intelligence. There have been many transitions toward greater complexity, none of them obvious: prokaryotic to eukaryotic unicellular creatures (and nothing more for 3 billion years!), unicellular to multicellular, sexual reproduction, mammals, intelligent mammals, Edge.org . . . Play the movie differently and we wouldn't be here.

As we look at planet Earth and the factors that enabled us to be here, we quickly realize that our planet is very special. Here's a short list: the long-term existence of a protective and oxygen-rich atmosphere; Earth's axial tilt, stabilized by a single large moon; the ozone layer and the magnetic field, which jointly protect surface creatures from lethal cosmic radiation; plate tectonics, which regulates the levels of carbon dioxide and keeps the global temperature stable; the fact that our sun is a smallish, fairly stable star not too prone to releasing huge plasma burps. Consequently, it's rather naïve to expect life—at the complexity level that exists here—to be ubiquitous across the universe.

A further point: Even if there is intelligent life elsewhere—and, of course, we can't rule that out (science is much better at finding things that exist than at ruling out things that don't)—it will be so remote that for all practical purposes we are alone. Even if SETI finds evidence of other cosmic intelligences, we are not going to initiate an intense collaboration. And if we are alone, and alone are aware of what it means to be alive and of the importance of remaining alive, we gain a new kind of cosmic centrality, very different and much more meaningful than the religion-inspired one of pre-Copernican days, when Earth was the center of Creation. We matter because we are rare and we know it.

The joint realization that we live in a remarkable cosmic cocoon and can create languages and rocket ships in an otherwise apparently dumb universe ought to be transformative. Until we find other self-aware intelligences, we are how the universe thinks. We might as well start enjoying one another's company.

THE MEDIOCRITY PRINCIPLE

P. Z. MYERS
Biologist, University of Minnesota; blogger, Pharyngula

As someone who just spent a term teaching freshman introductory biology and will be doing it again in the coming months, I have to say that the first thing that leaped to my mind as an essential skill everyone should have was algebra. And elementary probability and statistics. That sure would make my life easier, anyway; there's something terribly depressing about seeing bright students tripped up by a basic math skill they should have mastered in grade school.

But that isn't enough. Elementary math skills are an essential tool we ought to be able to take for granted in a scientific and technological society. What *idea* should people grasp to better understand their place in the universe?

I'm going to recommend the mediocrity principle. It's fundamental to science and it's also one of the most contentious, difficult concepts for many people to grasp. And opposition to the mediocrity principle is one of the major linchpins of religion and creationism and jingoism and failed social policies. There are a lot of cognitive ills that would be neatly wrapped up and easily disposed of if only everyone understood this one simple idea.

The mediocrity principle simply states that you aren't special. The universe does not revolve around you; this planet isn't privileged in any unique way; your country is not the perfect product of divine destiny; your existence isn't the product of directed, intentional fate; and that tuna sandwich you had for lunch was not plotting to give you indigestion. Most of what happens in the world is just a consequence of natural, universal laws—laws that apply everywhere and to everything, with no special exemptions or amplifications for your benefit—given variety by the input of chance. Everything that you as a human being consider cosmically important is an accident. The rules of inheritance and the nature of biology meant that when your parents had a baby, it was anatomically human and mostly fully functional physiologically, but the unique combination of traits that make you male or female, tall or short, brown-eyed or blue-eyed, were the result of a chance shuffle of genetic attributes during meiosis, a few random mutations, and the luck of the draw in the grand sperm race at fertilization.

Don't feel bad about that, though; it's not just you. The stars themselves form as a result of the properties of atoms, the specific features of each star set by the chance distribution of ripples of condensation through clouds of dust and gas. Our sun wasn't required to be where it is, with the luminosity it has; it just happens to be there, and our existence follows from this opportunity. Our species itself is partly shaped by the force of our environment through selection and partly by fluctuations of chance. If humans had gone extinct a hundred thousand years ago, the world would go on turning, life would go on thriving, and some other species would be prospering in our place—and most likely not by following the same intelligence-driven, technological path we did.

And that's OK—if you understand the mediocrity principle.

The reason this principle is so essential to science is that it's the beginning of understanding how we came to be here and how everything works. We look for general principles that apply to the universe as a whole first, and those explain much of the story; and then we look for the quirks and exceptions that led to the details. It's a strategy that succeeds and is useful in gaining a deeper knowledge. Starting with a presumption that a subject of interest represents a violation of the properties of the universe, that it was poofed uniquely into existence with a specific purpose, and that the conditions of its existence can no longer apply, means that you have leaped to an unfounded and unusual explanation with no legitimate reason. What the mediocrity principle tells us is that our state is not the product of intent, that the universe lacks both malice and benevolence, but that everything does follow rules—and that grasping those rules should be the goal of science.

THE POINTLESS UNIVERSE

SEAN CARROLL
Theoretical physicist, Caltech; author, From Eternity to Here: The Quest for the Ultimate Theory of Time

The world consists of things, which obey rules. If you keep asking "why" questions about what happens in the universe, you ultimately reach the answer "because of the state of the universe and the laws of nature."

This isn't an obvious way for people to think. Looking at the universe through our anthropocentric eyes, we can't help but view things in terms of causes, purposes, and natural ways of being. In ancient Greece, Plato and Aristotle saw the world teleologically—rain falls because water wants to be lower than air; animals (and slaves) are naturally subservient to human citizens.

From the start, there were skeptics. Democritus and Lucretius were early naturalists who urged us to think in terms of matter obeying rules rather than chasing final causes and serving underlying purposes. But it wasn't until our understanding of physics was advanced by thinkers such as Avicenna, Galileo, and Newton that it became reasonable to conceive of the universe evolving under its own power, free of guidance and support from anything beyond itself.

Theologians sometimes invoke "sustaining the world" as a function of God. But we know better; the world doesn't need to be sustained, it can simply be. Pierre-Simon Laplace articulated the very specific kind of rule that the world obeys: If we specify the complete state of the universe (or any isolated part of it) at some particular instant, the laws of physics tell us what its state will be at the very next moment. Applying those laws again, we can figure out what it will be a moment later. And so on, until (in principle, obviously) we can build up a complete history of the universe. This is not a universe that is advancing toward a goal; it is one that is caught in the iron grip of an unbreakable pattern.

This view of the processes at the heart of the physical world has important consequences for how we come to terms with the social world. Human beings like to insist that there are reasons why things happen. The death of a child, the crash of an airplane, or a random shooting must be explained in terms of the workings of a hidden plan. When Pat Robertson suggested that Hurricane Katrina was caused in part by God's anger at America's failing morals, he was attempting to provide an explanatory context for a seemingly inexplicable event.

Nature teaches us otherwise. Things happen because the laws of nature say they will—because they are the consequences of the state of the universe and the path of its evolution. Life on Earth doesn't arise in fulfillment of a grand scheme but as a by-product of the increase of entropy in an environment very far from equilibrium. Our impressive brains don't develop because life is guided toward greater levels of complexity and intelligence but from the mechanical interactions between genes, organisms, and their surroundings.

None of which is to say that life is devoid of purpose and meaning. Only that these are things we create, not things we discover out there in the fundamental architecture of the world. The world keeps happening, in accordance with its rules; it's up to us to make sense of it and give it value.

THE COPERNICAN PRINCIPLE

SAMUEL ARBESMAN
Applied mathematician; postdoctoral research fellow, Department of Health Care Policy, Harvard Medical School; affiliate, Institute for Quantitative Social Science, Harvard University

The scientist Nicolaus Copernicus recognized that Earth is not in any particularly privileged position in the solar system. This elegant fact can be extended to encompass a powerful idea, known as the Copernican Principle, which holds that we are not in a special or favorable place of any sort. By looking at the world in light of this principle, we can overcome certain preconceptions about ourselves and reexamine our relationship with the universe.

The Copernican Principle can be used in the traditional spatial sense, providing awareness of our sun's mediocre place in the suburbs of our galaxy and our galaxy's unremarkable place in the universe. And the Copernican Principle helps guide our understanding of the expanding universe, allowing us to see that anywhere in the cosmos one would perceive other galaxies moving away at rapid speeds, just as we see here on Earth. We are not anywhere special.

The Copernican Principle has also been extended to our temporal position by astrophysicist J. Richard Gott to help provide estimates for lifetimes of events, independent of additional information. As Gott elaborated, other than the fact that we are intelligent observers, there is no reason to believe we are in any way specially located in time. The Copernican Principle allows us to quantify our uncertainty and recognize that we are often neither at the beginning of things nor at the end. It allowed Gott to estimate correctly when the Berlin Wall would fall and has even provided meaningful numbers on the survival of humanity.

This principle can even anchor our location within the many orders of magnitude of our world: We are far smaller than most of the cosmos, far larger than most chemistry, far slower than much that occurs at subatomic scales, and far faster than geological and evolutionary processes. This principle leads us to study the successively larger and smaller orders of magnitude of our world, because we cannot assume that everything interesting is at the same scale as ourselves.

And yet despite this regimented approach to our mediocrity, we need not despair: As far as we know, we're the only species that recognizes its place in the universe. The paradox of the Copernican Principle is that by properly understanding our place, even if it be humbling, we can only then truly understand our particular circumstances. And when we do, we don't seem so insignificant after all.

WE ARE NOT ALONE IN THE UNIVERSE

J. CRAIG VENTER

Genome scientist; founder and president, J. Craig Venter Institute; author, A Life Decoded

I cannot imagine any single discovery that would have more impact on humanity than the discovery of life outside our solar system. There is a humancentric, Earthcentric view of life that permeates most cultural and societal thinking. Finding that there are multiple, perhaps millions, of origins of life and that life is ubiquitous throughout the universe will profoundly affect every human.

We live on a microbial planet. There are 1 million microbial cells per cubic centimeter of water in our oceans, lakes, and rivers; deep within the Earth's crust; and throughout our atmosphere. We have more than 100 trillion microbes on and in each of us. We have microbes that can withstand millions of rads of ionizing radiation or acids and bases so strong they would dissolve our skin. Microbes grow in ice, and microbes grow and thrive at temperatures exceeding 100 C°. We have life that lives on carbon dioxide, on methane, on sulfur, on sugar. We have sent trillions of bacteria into space over the last few billion years, and we have long exchanged material with Mars, so it would be very surprising if we do not find evidence of microbial life in our solar system, particularly on Mars.

The recent discoveries by Dimitar Sasselov and colleagues of numerous Earth and super-Earth-like planets outside our solar system, including water worlds, greatly increases the probability of finding life. Sasselov estimates that there are approximately a hundred thousand Earths and super-Earths within our own galaxy. The universe is young, so wherever we find microbial life, there will be intelligent life in the future.

Expanding our scientific reach farther into the skies will change us forever.

MICROBES RUN THE WORLD

STEWART BRAND
Founder, Whole Earth Catalog; cofounder, the WELL; cofounder, Global Business Network; author, Whole Earth Discipline

"Microbes run the world." That opening sentence of the National Research Council's *The New Science of Metagenomics* sounds reveille for a new way of understanding biology and maybe of understanding society as well.

The breakthrough was the shotgun sequencing of DNA, the same technology that gave us the human genome years ahead of schedule. Starting in 2003, Craig Venter and others began sequencing large populations of bacteria. The thousands of new genes they found (double the total previously discovered) showed what proteins the genes would generate and therefore what function they had, and that began to reveal what the teeming bacteria were really up to. This "meta"-genomics revolutionized microbiology, and that revolution will reverberate through the rest of biology for decades.

Microbes make up 80 percent of all biomass, says microbiologist Carl Woese. In one-fifth of a teaspoon of seawater, there are a million bacteria (and 10 million viruses), Craig Venter says, adding, "If you don't like bacteria, you're on the wrong planet. This is the planet of the bacteria." That means that most of the planet's living metabolism is microbial. When James Lovelock was trying to figure out where the gases come from that make the Earth's atmosphere such an artifact of life (the Gaia hypothesis), it was microbiologist Lynn Margulis who had the answer for him. Microbes run our atmosphere. They also run much of our body. The human microbiome in our gut, mouth, skin, and elsewhere, harbors three thousand kinds of bacteria with 3 million distinct genes. (Our own cells struggle by on only eighteen thousand genes or so.) New research is showing that our microbes-on-board drive our immune systems and important parts of our digestion.

Microbial evolution, which has been going on for more than 3.6 billion years, is profoundly different from what we think of as standard Darwinian evolution, where genes have to pass down generations to work slowly through the selection filter. Bacteria swap genes promiscuously within generations. They have three different mechanisms for this "horizontal gene transfer" among wildly different kinds of bacteria, and thus they evolve constantly and rapidly. Since they pass the opportunistically acquired genes on to their offspring, what they do on an hourly basis looks suspiciously Lamarckian—the inheritance of acquired characteristics.

Such routinely transgenic microbes show that there's nothing new, special, or dangerous about engineered GM crops. Field biologists are realizing that the biosphere is looking like what some are calling a pangenome, an interconnected network of continuously circulated genes that is a superset of all the genes in all the strains of a species that form. Bioengineers in the new field of synthetic biology are working directly with the conveniently fungible genes of microbes.

This biotech century will be microbe-enhanced and maybe microbe-inspired. Social Darwinism turned out to be a bankrupt idea. The term "cultural evolution" never meant much, because the fluidity of memes and influences in society bears no relation to the turgid conservatism of standard Darwinian evolution. But "social microbialism" might mean something as we continue to explore the fluidity of traits and the vast ingenuity of mechanisms among microbes—quorum sensing, biofilms, metabolic bucket brigades, "lifestyle genes," and the like. Confronting a difficult problem, we might fruitfully ask, "What would a microbe do?"

THE DOUBLE-BLIND CONTROL EXPERIMENT

RICHARD DAWKINS

Evolutionary zoologist, University of Oxford; author, The Greatest Show on Earth: The Evidence for Evolution

Not all concepts wielded by professional scientists would improve everybody's cognitive toolkit. We are here not looking for tools with which research scientists might benefit their science. We are looking for tools to help nonscientists understand science better and equip them to make better judgments throughout their lives.

Why do half of all Americans believe in ghosts, three-quarters believe in angels, a third believe in astrology, three-quarters believe in hell? Why do a quarter of all Americans believe that the president of the United States was born outside the country and is therefore ineligible to be president? Why do more than 40 percent of Americans think the universe began after the domestication of the dog?

Let's not give the defeatist answer and blame it all on stupidity. That's probably part of the story, but let's be optimistic and concentrate on something remediable: lack of training in how to think critically and how to discount personal opinion, prejudice, and anecdote in favor of evidence. I believe that the double-blind control experiment does double duty. It is more than just an excellent research tool. It also has educational, didactic value in teaching people how to think critically. My thesis is that you needn't actually do double-blind control experiments in order to experience an improvement in your cognitive toolkit. You need only understand the principle, grasp why it is necessary, and revel in its elegance.

If all schools taught their pupils how to do a double-blind control experiment, our cognitive toolkits would be improved in the following ways:

1. We would learn not to generalize from anecdotes.

2. We would learn how to assess the likelihood that an apparently important effect might have happened by chance alone.

3. We would learn how extremely difficult it is to eliminate subjective bias, and that subjective bias does not imply dishonesty or venality of any kind. This lesson goes deeper. It has the salutary effect of undermining respect for authority and respect for personal opinion.

4. We would learn not to be seduced by homeopaths and other quacks and charlatans, who would consequently be put out of business.

5. We would learn critical and skeptical habits of thought more generally, which not only would improve our cognitive toolkit but might save the world.

PROMOTING A SCIENTIFIC LIFESTYLE

MAX TEGMARK
Physicist, MIT; researcher, Precision Cosmology; scientific director, Foundational Questions Institute

I think the scientific concept that would most improve everybody's cognitive toolkit is "scientific concept."

Despite spectacular success in research, our global scientific community has been nothing short of a spectacular failure when it comes to educating the public. Haitians burned twelve "witches" in 2010. In the United States, recent polls show that 39 percent consider astrology scientific and 40 percent believe that our human species is less than ten thousand years old. If everyone understood the concept of "scientific concept," these percentages would be zero. Moreover, the world would be a better place, since people with a scientific lifestyle, basing their decisions on correct information, maximize their chances of success. By making rational buying and voting decisions, they also strengthen the scientific approach to decision making in companies, organizations, and governments.

Why have we scientists failed so miserably? I think the answers lie mainly in psychology, sociology, and economics.

A scientific lifestyle requires a scientific approach to both *gathering* information and *using* information, and both have their pitfalls. You're clearly more likely to make the right choice if you're aware of the full spectrum of arguments before making your mind up, yet there are many reasons why people don't get such complete information. Many lack access to it (3 percent of Afghans have access to the Internet, and in a 2010 poll 92 percent didn't know about the 9/11 attacks). Many are too swamped with obligations and distractions to seek it. Many seek information only from sources that confirm their preconceptions. Even for those who are online and uncensored, the most valuable information can be hard to find, buried in an unscientific media avalanche.

Then there's what we do with the information we have. The core of a scientific lifestyle is to change your mind when faced with information that disagrees with your views, avoiding intellectual inertia, yet many of us praise leaders who stubbornly stick to their views as "strong." The great physicist Richard Feynman hailed "distrust of experts" as a cornerstone of science, yet herd mentality and blind faith in authority figures is widespread. Logic forms the basis of scientific reasoning, yet wishful thinking, irrational fears, and other cognitive biases often dominate decisions.

What can we do to promote a scientific lifestyle?

The obvious answer is to improve education. In some countries, even the most rudimentary education would be a major improvement (less than half of all Pakistanis can read). By undercutting fundamentalism and intolerance, education would curtail violence and war. By empowering women, it would curb poverty and the population explosion.

However, even countries that offer everybody education can make major improvements. All too often, schools resemble museums, reflecting the past rather than shaping the future. The curriculum should shift from one watered down by consensus and lobbying to skills our century needs, for promoting relationships, health, contraception, time management, and critical thinking, and recognizing propaganda. For youngsters, learning a foreign language and typing should trump long division and writing cursive. In the Internet age, my own role as a classroom teacher has changed. I'm no longer needed as a conduit of information, which my students can simply download on their own; rather, my key role is inspiring a scientific lifestyle, curiosity, and the desire to learn.

Now let's get to the most interesting question: How can we *really* make a scientific lifestyle take root and flourish?

Reasonable people have been making similar arguments for better education since long before I was in diapers, yet instead of improving, education and adherence to a scientific lifestyle are arguably deteriorating in many countries, including the United States. Why? Clearly because there are powerful forces pushing in the opposite direction, and they are pushing more effectively. Corporations concerned that a better understanding of certain scientific issues would harm their profits have an incentive to muddy the waters, as do fringe religious groups concerned that questioning their pseudoscientific claims would erode their power.

So what can we do? The first thing we scientists need to do is get off our high horses, admit that our persuasive strategies have failed, and develop a better strategy. We have the advantage of having the better arguments, but the antiscientific coalition has the advantage of better funding.

However, and this is ironic, the antiscientific coalition is also more scientifically organized! If a company wants to change public opinion to increase their profits, it deploys scientific and highly effective marketing tools. What do people believe today? What do we want them to believe tomorrow? Which of their fears, insecurities, hopes, and other emotions can we take advantage of? What's the most cost-effective way of changing their minds? Plan a campaign. Launch. Done.

Is the message oversimplified or misleading? Does it unfairly discredit the competition? That's par for the course when marketing the latest smartphone or cigarette, so it would be naïve to think that the code of conduct should be any different when this coalition fights science.

Yet we scientists are often painfully naïve, deluding ourselves that just because we think we have the moral high ground, we can somehow defeat this corporate-fundamentalist coalition by using obsolete unscientific strategies. Based on what scientific argument will it make a hoot of difference if we grumble, "We won't stoop that low" and "People need to change" in faculty lunchrooms and recite statistics to journalists? We scientists have basically been saying "Tanks are unethical, so let's fight tanks with swords."

To teach people what a scientific concept is and how a scientific lifestyle will improve their lives, we need to go about it scientifically: We need new science advocacy organizations, which use all the same scientific marketing and fund-raising tools as the anti-scientific coalition. We'll need to use many of the tools that make scientists cringe, from ads and lobbying to focus groups that identify the most effective sound bites.

We won't need to stoop all the way down to intellectual dishonesty, however. Because in this battle, we have the most powerful weapon of all on our side: the facts.

EXPERIMENTATION

ROGER SCHANK
Psychologist and computer scientist, Engines for Education, Inc.; author, Making Minds Less Well Educated Than Our Own

Some scientific concepts have been so ruined by our education system that it is necessary to explain the ones that everyone thinks they know about and really don't.

We learn about experimentation in school. What we learn is that scientists conduct experiments, and in our high school labs if we copy exactly what they did, we will get the results they got. We learn about the experiments scientists do—usually about the physical and chemical properties of things—and we learn that they report their results in scientific journals. So, in effect, we learn that experimentation is boring, is something done by scientists, and has nothing to do with our daily lives.

And this is a problem. Experimentation is something done by everyone all the time. Babies experiment with what might be good to put in their mouths. Toddlers experiment with various behaviors to see what they can get away with. Teenagers experiment with sex, drugs, and rock and roll. But because people don't really see these things as experiments or as ways of collecting evidence in support or refutation of hypotheses, they don't learn to think about experimentation as something they do constantly and thus need to learn to do better.

Every time we take a prescription drug, we are conducting an experiment. But we don't carefully record the results after each dose, and we don't run controls, and we mix up the variables by not changing only one behavior at a time, so that when we suffer from side effects we can't figure out what might have been their true cause. We do the same with personal relationships: When they go wrong, we can't figure out why, because the conditions are different in each one.

Now, while it is difficult if not impossible to conduct controlled experiments in most aspects of our lives, it is possible to come to understand that we are indeed conducting an experiment when we take a new job, or try a new tactic in a game, or pick a school to attend—or when we try and figure out how someone is feeling or wonder why we ourselves feel as we do.

Every aspect of life is an experiment that can be better understood if it is perceived in that way. But because we don't recognize this, we fail to understand that we need to reason logically from evidence we gather, carefully consider the conditions under which our experiment has been conducted, and decide when and how we might run the experiment again with better results. The scientific activity that surrounds experimentation is about thinking clearly in the face of evidence obtained from the experiment. But people who don't see their actions as experiments and don't know how to reason carefully from data will continue to learn less well from their experiences than those who do.

Most of us, having learned the word "experiment" in the context of a boring ninth-grade science class, have long since learned to discount science and experimentation as irrelevant to our lives. If schools taught basic cognitive concepts, such as experimentation in the context of everyday experience, instead of concentrating on algebra as a way of teaching people how to reason, then people would be much more effective at thinking about politics, child raising, personal relationships, business, and every other aspect of their daily lives.

THE CONTROLLED EXPERIMENT

TIMO HANNAY
Managing director, Digital Science, Macmillan Publishers Ltd.

The scientific concept that most people would do well to understand and exploit is the one that almost defines science itself: the controlled experiment.

When they are required to make a decision, the instinctive response of most nonscientists is to introspect, or perhaps call a meeting. The scientific method dictates that wherever possible we should instead conduct a suitable controlled experiment. The superiority of the latter approach is demonstrated not only by the fact that science has uncovered so much about the world but also, and even more powerfully, by the fact that such a lot of it—the Copernican Principle, evolution by natural selection, general relativity, quantum mechanics—is so mind-bendingly counterintuitive. Our embrace of truth as defined by experiment (rather than by common sense, or consensus, or seniority, or revelation, or any other means) has, in effect, released us from the constraints of our innate preconceptions, prejudices, and lack of imagination. It has freed us to appreciate the universe in terms well beyond our abilities to derive by intuition alone.

What a shame, then, that experiments are by and large performed only by scientists. What if businesspeople and policy makers were to spend less time relying on instinct or partially informed debate and more time devising objective ways to identify the best answers? I think that would often lead to better decisions.

In some domains, this is already starting to happen. Online companies, such as Amazon and Google, don't anguish over how to design their Web sites. Instead, they conduct controlled experiments by showing different versions to different groups of users until they have iterated to an optimal solution. (And with the amount of traffic those sites receive, individual tests can be completed in seconds.) They are helped, of course, by the fact that the Web is particularly conducive to rapid data acquisition and product iteration. But they are helped even more by the fact that their leaders often have backgrounds in engineering or science and therefore adopt a scientific—which is to say, experimental—mind-set.

Government policies—from teaching methods in schools to prison sentencing to taxation —would also benefit from more use of controlled experiments. This is where many people start to get squeamish. To become the subject of an experiment in something as critical or controversial as our children's education or the incarceration of criminals feels like an affront to our sense of fairness and our strongly held belief in the right to be treated exactly the same as everybody else. After all, if there are separate experimental and control groups, then surely one of them must be losing out. Well, no, because we do not know in advance which group will be better off, which is precisely why we are conducting the experiment. Only when a potentially informative experiment is not conducted do true losers arise: all those future generations who stood to benefit from the results. The real reason people are uncomfortable is simply that we're not used to seeing experiments conducted in these domains. After all, we willingly accept them in the much more serious context of clinical trials, which are literally matters of life and death.

Of course, experiments are not a panacea. They will not tell us, for example, whether an accused person is innocent or guilty. Moreover, experimental results are often inconclusive. In such circumstances, a scientist can shrug his shoulders and say that he is still unsure, but a businessperson or lawmaker will often have no such luxury and may be forced to make a decision anyway. Yet none of this takes away from the fact that the controlled experiment is the best method yet devised to reveal truths about the world, and we should use them wherever they can be sensibly applied.

GEDANKENEXPERIMENT

GINO SEGRE
Professor of physics at the University of Pennsylvania; author, Ordinary Geniuses: Max Delbrück, George Gamow, and the Origins of Genomics and Big Bang Cosmology

The notion of a gedankenexperiment, or thought experiment, has been integral to the theoretical physics toolkit ever since that discipline came into existence. It involves setting up an imagined piece of apparatus and running a simple experiment with it in your mind, for the purpose of proving or disproving a hypothesis. In many cases, a gedankenexperiment is the only approach. An actual experiment to examine retrieval of information falling into a black hole cannot be carried out.

The notion was particularly important during the development of quantum mechanics, when legendary gedankenexperiments were conducted by the likes of Niels Bohr and Albert Einstein to test such novel ideas as the uncertainty principle and wave-particle duality. Examples, like that of "Schrödinger's cat," have even come into the popular lexicon. Is the cat simultaneously dead and alive? Others, particularly the classic double slit through which a particle/wave passes, were part of the first attempt to understand quantum mechanics and have remained as tools for understanding its meaning.

However, the subject need not be an esoteric one for a gedankenexperiment to be fruitful. My own favorite is Galileo's proof that, contrary to Aristotle's view, objects of different mass fall in a vacuum with the same acceleration. One might think that a real experiment needs to be conducted to test that hypothesis, but Galileo simply asked us to consider a large and a small stone tied together by a very light string. If Aristotle was right, the large stone should speed up the smaller one, and the smaller one retard the larger one, if they fall at different rates. However, if you let the string length approach zero, you have a single object with a mass equal to the sum of their masses, and hence it should fall at a higher rate than either. This is nonsensical. The conclusion is that all objects fall in a vacuum at the same rate.

Consciously or unconsciously, we carry out gedankenexperiments of one sort or another in our everyday life and are even trained to perform them in a variety of disciplines, but it would be useful to have a greater awareness of how they are conducted and how they can be positively applied. We could ask, when confronted with a puzzling situation, "How can I set up a gedankenexperiment to resolve the issue?" Perhaps our financial, political, and military experts would benefit from following such a tactic—and arrive at happier outcomes.

THE PESSIMISTIC META-INDUCTION FROM THE HISTORY OF SCIENCE

KATHRYN SCHULZ
Journalist; author, Being Wrong: Adventures in the Margin of Error

OK, OK: It's a terrible phrase. In my defense, I didn't coin it; philosophers of science have been kicking it around for a while. But if "the pessimistic meta-induction from the history of science" is cumbersome to say and difficult to remember, it is also a great idea. In fact, as the "meta" part suggests, it's the kind of idea that puts all other ideas into perspective.

Here's the gist: Because so many scientific theories from bygone eras have turned out to be wrong, we must assume that most of today's theories will eventually prove incorrect as well. And what goes for science goes in general. Politics, economics, technology, law, religion, medicine, child rearing, education: No matter the domain of life, one generation's verities so often become the next generation's falsehoods that we might as well have a pessimistic meta-induction from the history of everything.

Good scientists understand this. They recognize that they are part of a long process of approximation. They know they are constructing models rather than revealing reality. They are comfortable working under conditions of uncertainty—not just the local uncertainty of "Will this data bear out my hypothesis?" but the sweeping uncertainty of simultaneously pursuing and being cut off from absolute truth.

The rest of us, by contrast, often engage in a kind of tacit chronological exceptionalism. Unlike all those suckers who fell for the flat Earth or the geocentric universe or cold fusion, we ourselves have the great good luck to be alive during the very apex of accurate human thought. The literary critic Harry Levin put this nicely: "The habit of equating one's age with the apogee of civilization, one's town with the hub of the universe, one's horizons with the limits of human awareness, is paradoxically widespread." At best, we nurture the fantasy that knowledge is always cumulative and therefore concede that future eras will know more than we do. But we ignore or resist the fact that knowledge collapses as often as it accretes, that our own most cherished beliefs might appear patently false to posterity.

That fact is the essence of the meta-induction—and yet, despite its name, this idea is not pessimistic. Or rather, it is pessimistic only if you hate being wrong. If, by contrast, you think that uncovering your mistakes is one of the best ways to revise and improve your understanding of the world, then this is actually a highly optimistic insight.

The idea behind the meta-induction is that all of our theories are fundamentally provisional and quite possibly wrong. If we can add that idea to our cognitive toolkit, we will be better able to listen with curiosity and empathy to those whose theories contradict our own. We will be better able to pay attention to counterevidence—those anomalous bits of data that make our picture of the world a little weirder, more mysterious, less clean, less done. And we will be able to hold our own beliefs a bit more humbly, in the happy knowledge that better ideas are almost certainly on the way.

EACH OF US IS ORDINARY, YET ONE OF A KIND

SAMUEL BARONDES

Director of the Center for Neurobiology & Psychiatry at the University of California–San Francisco; author, Making Sense of People: Decoding the Mysteries of Personality

Each of us is ordinary, yet one of a kind.

Each of us is standard issue, conceived by the union of two germ cells, nurtured in a womb, and equipped with a developmental program that guides our further maturation and eventual decline.

Each of us is also unique, the possessor of a particular selection of gene variants from the collective human genome and immersed in a particular family, culture, era, and peer group. With inborn tools for adaptation to the circumstances of our personal world, we keep building our own ways of being and the sense of who we are.

This dual view of each of us, as both run-of-the-mill and special, has been so well established by biologists and behavioral scientists that it may now seem self-evident. But it still deserves conscious attention as a specific cognitive chunk, because it has such important implications. Recognizing how much we share with others promotes compassion, humility, respect, and brotherhood. Recognizing that we are each unique promotes pride, self-development, creativity, and achievement.

Embracing these two aspects of our personal reality can enrich our daily experience. It allows us to simultaneously enjoy the comfort of being ordinary and the excitement of being one of a kind.

NEXUS CAUSALITY, MORAL WARFARE, AND MISATTRIBUTION ARBITRAGE

JOHN TOOBY

Founder of the field of evolutionary psychology; codirector, University of California–Santa Barbara's Center for Evolutionary Psychology

We could become far more intelligent than we are by adding to our stock of concepts and forcing ourselves to use them even when we don't like what they are telling us. This will be nearly always, because they generally tell us that our self-evidently superior selves and in-groups are error-besotted. We all start from radical ignorance in a world that is endlessly strange, vast, complex, intricate, and surprising. Deliverance from ignorance lies in good concepts—inference fountains that geyser out insights that organize and increase the scope of our understanding. We are drawn to them by the fascination of the discoveries they afford, but we resist using them well and freely, because they would reveal too many of our apparent achievements to be embarrassing or tragic failures. Those of us who are nonmythical lack the spine that Oedipus had—the obsidian resolve that drove him to piece together shattering realizations despite portents warning him off. Because of our weakness, "to see what is in front of one's nose needs a constant struggle," as Orwell says. So why struggle? Better instead to have one's nose and what lies beyond shift out of focus—to make oneself hysterically blind as convenience dictates, rather than risk ending up like Oedipus, literally blinding oneself in horror at the harvest of an exhausting, successful struggle to discover what is true.

Alternatively, even modest individual-level improvements in our conceptual toolkit can have transformative effects on our collective intelligence, promoting incandescent intellectual chain reactions among multitudes of interacting individuals. If this promise of intelligence-amplification through conceptual tools seems like hyperbole, consider that the least inspired modern engineer, equipped with the conceptual tools of calculus, can understand, plan, and build things far beyond what Leonardo or the mathematics-revering Plato could have achieved without it. We owe a lot to the infinitesimal, Newton's counterintuitive conceptual hack—something greater than zero but less than any finite magnitude. Much simpler conceptual innovations than calculus have had even more far-reaching effects—the experiment (a danger to authority), zero, entropy, Boyle's atom, mathematical proof, natural selection, randomness, particulate inheritance, Dalton's element, distribution, formal logic, culture, Shannon's definition of information, the quantum . . .

Here are three simple conceptual tools that might help us see in front of our noses: *nexus causality*, *moral warfare*, and *misattribution arbitrage*. Causality itself is an evolved conceptual tool that simplifies, schematizes, and focuses our representation of situations. This cognitive machinery guides us to think in terms of *the* cause—of an outcome's having a single cause. Yet for enlarged understanding, it is more accurate to represent outcomes as caused by an intersection, or nexus, of factors (including the absence of precluding conditions). In *War and Peace*, Tolstoy asks, "When an apple ripens and falls, why does it fall? Because of its attraction to the earth, because its stem withers, because it is dried by the sun, because it grows heavier, because the wind shakes it . . . ?" With little effort, any modern scientist could extend Tolstoy's list endlessly. We evolved, however, as cognitively improvisational tool users, dependent on identifying actions we could take that would lead to immediate payoffs. So our minds evolved to represent situations in a way that highlighted the element in the nexus that we could manipulate to bring about a favored outcome. Elements in the situation that remained stable and that we could not change (like gravity or human nature) were left out of our representation of causes. Similarly, variable factors in the nexus (like the wind blowing), which we could not control but which predicted an outcome (the apple falling), were also useful to represent as causes, to prepare ourselves to exploit opportunities or avoid dangers. So the reality of the causal nexus is cognitively ignored in favor of the cartoon of single causes. While useful for a forager, this machinery impoverishes our scientific understanding, rendering discussions (whether elite, scientific, or public) of the "causes"—of cancer, war, violence, mental disorders, infidelity, unemployment, climate, poverty, and so on—ridiculous.

Similarly, as players of evolved social games, we are designed to represent others' behavior and associated outcomes as caused by free will (by *intentions*). That is, we evolved to view "man," as Aristotle put it, as "the originator of his own actions." Given an outcome we dislike, we ignore the nexus and trace "the" causal chain back to a person. We typically represent the backward chain as ending in—and the outcome as originating in—the person. Locating the "cause" (blame) in one or more persons allows us to punitively motivate others to avoid causing outcomes we don't like (or to incentivize outcomes we do like). More despicable, if something happens that many regard as a bad outcome, this gives us the opportunity to sift through the causal nexus for the one thread that colorably leads back to our rivals (where the blame obviously lies). Lamentably, much of our species' moral psychology evolved for moral warfare, a ruthless zero-sum game. Offensive play typically involves recruiting others to disadvantage or eliminating our rivals by publicly sourcing them as the cause of bad outcomes.

Defensive play involves giving our rivals no ammunition to mobilize others against us.

The moral game of blame attribution is only one subtype of misattribution arbitrage. For example, epidemiologists estimate that it was not until 1905 that you were better off going to a physician. (Ignaz Semelweiss noticed that doctors doubled the mortality rate of mothers at delivery.) The role of the physician predated its rational function for thousands of years, so why were there physicians? Economists, forecasters, and professional portfolio managers typically do no better than chance, yet command immense salaries for their services. Food prices are driven up to starvation levels in underdeveloped countries, based on climate models that cannot successfully retrodict known climate history. Liability lawyers win huge sums for plaintiffs who get diseases at no higher rates than others not exposed to "the" supposed cause. What is going on? The complexity and noise permeating any real causal nexus generates a fog of uncertainty. Slight biases in causal attribution or in blameworthiness (e.g., sins of commission are worse than sins of omission) allow a stable niche for extracting undeserved credit or targeting undeserved blame. If the patient recovers, it was due to my heroic efforts; if not, the underlying disease was too severe. If it weren't for my macroeconomic policy, the economy would be even worse. The abandonment of moral warfare and a wider appreciation of nexus causality and misattribution arbitrage would help us all shed at least some of the destructive delusions that cost humanity so much.

SELF-SERVING BIAS

DAVID G. MYERS
Social psychologist, Hope College; author, A Friendly Letter to Skeptics and Atheists

Most of us have a good reputation with ourselves. That's the gist of a sometimes amusing and frequently perilous phenomenon that social psychologists call self-serving bias.

Accepting more responsibility for success than for failure, for good deeds than for bad.

In experiments, people readily accept credit when told they have succeeded, attributing it to their ability and effort. Yet they attribute failure to such external factors as bad luck or the problem's "impossibility." When we win at Scrabble, it's because of our verbal dexterity. When we lose, it's because "I was stuck with a Q but no U." Self-serving attributions have been observed with athletes (after victory or defeat), students (after high or low exam grades), drivers (after accidents), and managers (after profits or losses). The question "What have I done to deserve this?" is one we ask of our troubles, not our successes.

The better-than-average phenomenon: How do I love me? Let me count the ways.

It's not just in Lake Wobegon that all the children are above average. In one College Board survey of 829,000 high-school seniors, 0 percent rated themselves below average in "ability to get along with others," 60 percent rated themselves in the top 10 percent, and 25 percent rated themselves in the top 1 percent. Compared with our average peer, most of us fancy ourselves as more intelligent, better-looking, less prejudiced, more ethical, healthier, and likely to live longer—a phenomenon recognized in Freud's joke about the man who told his wife, "If one of us should die, I shall move to Paris."

In everyday life, more than nine in ten drivers are above-average drivers, or so they presume. In surveys of college faculty, 90 percent or more have rated themselves as superior to their average colleague (which naturally leads to some envy and disgruntlement when one's talents are underappreciated). When husbands and wives estimate what percent of the housework they contribute, or when work-team members estimate their contributions, their self-estimates routinely sum to more than 100 percent.

Studies of self-serving bias and its cousins—illusory optimism, self-justification, and in-group bias—remind us of what literature and religion have taught: Pride often goes before a fall. Perceiving ourselves and our group favorably protects us against depression, buffers stress, and sustains our hopes. But it does so at the cost of marital discord, bargaining impasses, condescending prejudice, national hubris, and war. Being mindful of self-serving bias beckons us not to false modesty but to a humility that affirms our genuine talents and virtues and likewise those of others.

COGNITIVE HUMILITY

GARY MARCUS
Director, Center for Child Language, New York University; author, Kluge: The Haphazard Evolution of the Human Mind

Hamlet may have said that human beings are noble in reason and infinite in faculties, but in reality—as four decades of experiments in cognitive psychology have shown—our minds are finite and far from noble. Knowing their limits can help us to become better reasoners.

Almost all of those limits start with a peculiar fact about human memory: Although we are pretty good at storing information in our brains, we are pretty poor at retrieving it. We can recognize photos from our high school yearbooks decades later, yet find it impossible to remember what we had for breakfast yesterday. Faulty memories have been known to lead to erroneous eyewitness testimony (and false imprisonment), to marital friction (in the form of overlooked anniversaries), and even death (skydivers, for example, have been known to forget to pull their rip cords, accounting by one estimate for approximately 6 percent of diving fatalities).

Computer memory is much better than human memory because early computer scientists discovered a trick that evolution never did: organizing information by assigning every memory to a master map in which each bit of information to be stored is assigned a uniquely identifiable location in the computer's memory vaults. Human beings, by contrast, appear to lack such master memory maps and retrieve information in far more haphazard fashion, by using clues (or cues) to what's being looked for. In consequence, our memories cannot be searched as systematically or as reliably as that of a computer (or Internet database). Instead, human memories are deeply subject to context. Scuba divers, for example, are better at remembering the words they study underwater when they are tested underwater rather than on land, even if the words have nothing to do with the sea.

Sometimes this sensitivity to context is useful. We are better able to remember what we know about cooking when we're in the kitchen than when we're, say, skiing. But it also comes at a cost: When we need to remember something in a situation other than the one in which it was stored, the memory is often hard to retrieve. One of the biggest challenges in education, for example, is to get children to apply what they learn in school to real-world situations. Perhaps the most dire consequence is that human beings tend to be better at remembering evidence consistent with their beliefs than evidence that contradicts those beliefs. When two people disagree, it is often because their prior beliefs lead them to remember (or focus on) different bits of evidence. To consider something well, of course, is to evaluate both sides of an argument, but unless we also go the extra mile of deliberately forcing ourselves to consider alternatives—which doesn't come naturally—we're more prone to recall evidence consistent with a belief than inconsistent with it.

Overcoming this mental weakness (known as confirmation bias) is a lifelong struggle; recognizing that we all suffer from it is an important first step. We can try to work around it, compensating for our inborn tendencies toward self-serving and biased recollection by disciplining ourselves to consider not just the data that might fit with our own beliefs but also the data that might lead other people to have beliefs different from ours.

TECHNOLOGIES HAVE BIASES

DOUGLAS RUSHKOFF

Media theorist; documentary writer; author, Program or Be Programmed: Ten Commands for a Digital Age

People like to think of technologies and media as neutral and that only their use or content determines their effect. Guns don't kill people, after all; people kill people. But guns are much more biased toward killing people than, say, pillows—even though many a pillow has been utilized to smother an aging relative or adulterous spouse.

Our widespread inability to recognize or even acknowledge the biases of the technologies we use renders us incapable of gaining any real agency through them. We accept our iPads, Facebook accounts, and automobiles at face value—as preexisting conditions—rather than as tools with embedded biases.

Marshall McLuhan exhorted us to recognize that our media affect us beyond whatever content is being transmitted through them. And while his message was itself garbled by the media through which he expressed it (the medium is the what?), it is true enough to be generalized to all technology. We are free to use any car we like to get to work—gasoline-, diesel-, electric-, or hydrogen-powered—and this sense of choice blinds us to the fundamental bias of the automobile toward distance, commuting, suburbs, and energy consumption.

Likewise, soft technologies, from central currency to psychotherapy, are biased in their construction as much as in their implementation. No matter how we spend U.S. dollars, we are nonetheless fortifying banking and the centralization of capital. Put a psychotherapist on his own couch and a patient in the chair and the therapist will begin to exhibit treatable pathologies. It's set up that way, just as Facebook is set up to make us think of ourselves in terms of our "likes" and an iPad is set up to make us start paying for media and stop producing them ourselves.

If the concept that technologies have biases were to become common knowledge, we could implement them consciously and purposefully. If we don't bring this concept into general awareness, our technologies and their effects will continue to threaten and confound us.

BIAS IS THE NOSE FOR THE STORY

GERALD SMALLBERG
Practicing neurologist, New York City; playwright, off-off-Broadway: Charter Members, The Gold Ring

The exponential explosion of information and its accessibility make our ability to gauge its truthfulness not only more important but also more difficult. Information has importance in proportion to its relevance and meaning. Its ultimate value is how we use it to make decisions and put it in a framework of knowledge.

Our perceptions are crucial in appreciating truth. However, we do not apprehend objective reality. Perception is based on recognition and interpretation of sensory stimuli derived from patterns of electrical impulses. From this data, the brain creates analogs and models that simulate tangible, concrete objects in the real world. Experience, though, colors and influences all of our perceptions by anticipating and predicting everything we encounter. It is the reason Goethe advised that "one must ask children and birds how cherries and strawberries taste." This preferential set of intuitions, feelings, and ideas—less poetically characterized by the term "bias"—poses a challenge to our ability to weigh evidence accurately to arrive at truth. Bias is the thumb that experience puts on the scale.

Our brains evolved having to make the right bet with limited information. Fortune, it has been said, favors the prepared mind. Bias, in the form of expectation, inclination, and anticipatory hunches, helped load the dice in our favor and for that reason is hardwired into our thinking. Bias is an intuition—a sensitivity, a receptiveness—that acts as a lens or filter on all our perceptions.

"If the doors of perception were cleansed," William Blake said, "every thing would appear to man as it is, infinite." But without our biases to focus our attention, we would be lost in that endless and limitless expanse. We have at our disposal an immeasurable assortment of biases, and their combination in each of us is as unique as a fingerprint. These biases mediate between our intellect and emotions to help congeal perception into opinion, judgment, category, metaphor, analogy, theory, and ideology, which frame how we see the world.

Bias is tentative. Bias adjusts as the facts change. Bias is a provisional hypothesis. Bias is normal.

Although bias is normal in the sense that it is a product of how we select and perceive information, its influence on our thinking cannot be ignored. Medical science has long been aware of the inherent bias that occurs in collecting and analyzing clinical data. The double-blind, randomized, controlled study, the gold standard of clinical design, was developed in an attempt to nullify its influence.

We live in the world, however, not in a laboratory, and bias cannot be eliminated. Bias, critically utilized, sharpens the collection of data by knowing when to look, where to look, and how to look. It is fundamental to both inductive and deductive reasoning. Darwin didn't collect his data randomly or disinterestedly to formulate the theory of evolution by natural selection. Bias is the nose for the story.

Truth needs continually to be validated against all evidence that challenges it fairly and honestly. Science, with its formal methodology of experimentation and the reproducibility of its findings, is available to anyone who plays by its rules. No ideology, religion, culture, or civilization is awarded special privileges or rights. The truth that survives this ordeal has another burden to bear. Like the words in a multidimensional crossword puzzle, it has to fit together with all the other pieces already in place. The better and more elaborate the fit, the more certain the truth. Science permits no exceptions. It is inexorably revisionary, learning from its mistakes, erasing and rewriting even its most sacred texts, until the puzzle is complete.

CONTROL YOUR SPOTLIGHT

JONAH LEHRER
Contributing editor, Wired *magazine; author,* How We Decide

In the late 1960s, the psychologist Walter Mischel began a simple experiment with four-year-old children. He invited the kids into a tiny room containing a desk and a chair and asked them to pick a treat from a tray of marshmallows, cookies, and pretzel sticks. Mischel then made the four-year-olds an offer: They could either eat one treat right away or, if they were willing to wait while he stepped out for a few minutes, they could have two treats when he returned. Not surprisingly, nearly every kid chose to wait.

At the time, psychologists assumed that the ability to delay gratification in order to get that second marshmallow or cookie depended on willpower. Some people simply had more willpower than others, which allowed them to resist tempting sweets and save money for retirement. However, after watching hundreds of kids participate in the marshmallow experiment, Mischel concluded that this standard model was wrong. He came to realize that willpower was inherently weak and that children who tried to postpone the treat—gritting their teeth in the face of temptation—soon lost the battle, often within thirty seconds.

Instead, Mischel discovered something interesting when he studied the tiny percentage of kids who could successfully wait for the second treat. Without exception, these "high delayers" all relied on the same mental strategy: They found a way to keep themselves from thinking about the treat, directing their gaze away from the yummy marshmallow. Some covered their eyes or played hide-and-seek underneath the desks. Others sang songs from *Sesame Street*, or repeatedly tied their shoelaces, or pretended to take a nap. Their desire wasn't defeated, it was merely forgotten.

Mischel refers to this skill as the "strategic allocation of attention," and he argues that it's the skill underlying self-control. Too often, we assume that willpower is about having strong moral fiber. But that's wrong. Willpower is really about properly directing the spotlight of attention, learning how to control that short list of thoughts in working memory. It's about realizing that if we're thinking about the marshmallow, we're going to eat it, which is why we need to look away.

What's interesting is that this cognitive skill isn't just useful for dieters. It seems to be a core part of success in the real world. For instance, when Mischel followed up with the initial subjects thirteen years later—they were now high school seniors—he realized that their performance on the marshmallow task had been highly predictive on a vast range of metrics. Those kids who had struggled to wait at the age of four were also more likely to have behavioral problems, both in school and at home. They struggled in stressful situations, often had trouble paying attention, and found it difficult to maintain friendships. Most impressive, perhaps, were the academic numbers: The kids who could wait fifteen minutes for a marshmallow had an SAT score that was, on average, 210 points higher than that of the kids who could wait only thirty seconds.

These correlations demonstrate the importance of learning to strategically allocate our attention. When we properly control the spotlight, we can resist negative thoughts and dangerous temptations. We can walk away from fights and improve our odds against addiction. Our decisions are driven by the facts and feelings bouncing around the brain—the allocation of attention allows us to direct this haphazard process, as we consciously select the thoughts we want to think about.

Furthermore, this mental skill is getting more valuable. We

live, after all, in the age of information, which makes the ability to focus on the important information incredibly important. (Herbert Simon said it best: "A wealth of information creates a poverty of attention.") The brain is a bounded machine, and the world is a confusing place, full of data and distractions. Intelligence is the ability to parse the data so that it makes just a little bit more sense. Like willpower, this ability requires the strategic allocation of attention.

One final thought: In recent decades, psychology and neuroscience have severely eroded classical notions of free will. The unconscious mind, it turns out, is most of the mind. And yet, we can still control the spotlight of attention, focusing on those ideas that will help us succeed. In the end, this may be the only thing we can control. We don't have to look at the marshmallow.

THE FOCUSING ILLUSION

DANIEL KAHNEMAN

Professor emeritus of psychology and public affairs, Woodrow Wilson School, Princeton University; recipient, 2002 Nobel Memorial Prize in Economic Sciences

Education is an important determinant of income—one of the most important—but it is less important than most people think. If everyone had the same education, the inequality of income would be reduced by less than 10 percent. When you focus on education, you neglect the myriad other factors that determine income. The differences of income among people who have the same education are huge.

Income is an important determinant of people's satisfaction with their lives, but it is far less important than most people think. If everyone had the same income, the differences among people in life satisfaction would be reduced by less than 5 percent.

Income is even less important as a determinant of emotional happiness. Winning the lottery is a happy event, but the elation does not last. On average, individuals with high income are in a better mood than people with lower income, but the difference is about a third as large as most people expect. When you think of rich and poor people, your thoughts are inevitably focused on circumstances in which income is important. But happiness depends on other factors more than it depends on income.

Paraplegics are often unhappy, but they are not unhappy all the time, because they spend most of the time experiencing and thinking about things other than their disability. When we think of what it is like to be a paraplegic, or blind, or a lottery winner, or a resident of California, we focus on the distinctive aspects of each of these conditions. The mismatch in the allocation of attention between thinking about a life condition and actually living it is the cause of the focusing illusion.

Marketers exploit the focusing illusion. When people are induced to believe that they "must have" a good, they greatly exaggerate the difference that the good will make to the quality of their life. The focusing illusion is greater for some goods than for others, depending on the extent to which the goods attract continued attention over time. The focusing illusion is likely to be more significant for leather car seats than for books on tape.

Politicians are almost as good as marketers in causing people to exaggerate the importance of issues on which their attention is focused. People can be made to believe that school uniforms will significantly improve educational outcomes, or that health care reform will hugely change the quality of life in the United States—either for the better or for the worse. Health care reform will make a difference, but the difference will be smaller than it appears when you focus on it.

THE USELESSNESS OF CERTAINTY

CARLO ROVELLI
Physicist, Centre de Physique Théorique, Marseille, France; author, The First Scientist: Anaximander and His Legacy

There is a widely held notion that does plenty of damage: the notion of "scientifically proved." Nearly an oxymoron. The very foundation of science is to keep the door open to doubt. Precisely because we keep questioning everything, especially our own premises, we are always ready to improve our knowledge. Therefore a good scientist is never "certain." Lack of certainty is precisely what makes conclusions more reliable than the conclusions of those who are certain, because the good scientist will be ready to shift to a different point of view if better evidence or novel arguments emerge. Therefore certainty is not only something of no use but is also in fact damaging, if we value reliability.

Failure to appreciate the value of uncertainty is at the origin of much silliness in our society. Are we sure that the Earth is going to keep heating up if we don't do anything? Are we sure of the details of the current theory of evolution? Are we sure that modern medicine is always a better strategy than traditional ones? No, we are not, in any of these cases. But if, from this lack of certainty, we jump to the conviction that we had better not care about global heating, that there is no evolution and the world was created six thousand years ago, or that traditional medicine must be more effective than modern medicine—well, we are simply stupid. Still, many people do make these inferences, because the lack of certainty is perceived as a sign of weakness instead of being what it is—the first source of our knowledge.

Every knowledge, even the most solid, carries a margin of

uncertainty. (I am very sure what my own name is . . . but what if I just hit my head and got momentarily confused?) Knowledge itself is probabilistic in nature, a notion emphasized by some currents of philosophical pragmatism. A better understanding of the meaning of "probability"—and especially realizing that we don't need (and never possess) "scientifically proved" facts but only a sufficiently high degree of probability in order to make decisions—would improve everybody's conceptual toolkit.

UNCERTAINTY

LAWRENCE KRAUSS
Physicist, Foundation Professor, and director, Origins Project, Arizona State University; author, Quantum Man: Richard Feynman's Life in Science

The notion of uncertainty is perhaps the least well understood concept in science. In the public parlance, uncertainty is a bad thing, implying a lack of rigor and predictability. The fact that global warming estimates are uncertain, for example, has been used by many to argue against any action at the present time.

In fact, however, uncertainty is a central component of what makes science successful. Being able to quantify uncertainty and incorporate it into models is what makes science quantitative rather than qualitative. Indeed, no number, no measurement, no observable in science is exact. Quoting numbers without attaching an uncertainty to them implies that they have, in essence, no meaning.

One of the things that makes uncertainty difficult for members of the public to appreciate is that the significance of uncertainty is relative. Take, for example, the distance between Earth and the sun: 1.49597 x 108 km., as measured at one point during the year. This seems relatively precise; after all, using six significant digits means I know the distance to an accuracy of one part in a million or so. However, if the next digit is uncertain, that means the uncertainty in knowing the precise Earth-sun distance is larger than the distance between New York and Chicago!

Whether or not the quoted number is "precise" therefore depends on what I'm intending to do with it. If I care only about what minute the sun will rise tomorrow, then the number quoted here is fine. If I want to send a satellite to orbit just above the sun, however, then I would need to know distances more accurately.

This is why uncertainty is so important. Until we can quantify the uncertainty in our statements and our predictions, we have little idea of their power or significance. So, too, in the public sphere. Public policy performed in the absence of understanding quantitative uncertainties, or even absent understanding the difficulty of obtaining reliable estimates of uncertainties, usually means bad public policy.

A SENSE OF PROPORTION ABOUT FEAR OF THE UNKNOWN

AUBREY DE GREY

Gerontologist; chief science officer, SENS Foundation; coauthor (with Michael Rae), Ending Aging

Einstein ranks extremely high not only among the all-time practitioners of science but also among the producers of aphorisms that place science in its real-world context. One of my favorites is "If we knew what we were doing, it wouldn't be called research." This disarming comment, like so many of the best quotes by experts in any field, embodies a subtle mix of sympathy and disdain for the difficulties the great unwashed have in appreciating what those experts do.

One of the foremost challenges facing scientists today is to communicate the management of uncertainty. The public knows that experts are, well, expert—that they know more than anyone else about the issue at hand. What is evidently far harder for most people to grasp is that "more than anyone else" does not mean "everything"—and especially that given the possession of only partial knowledge, experts must also be expert at figuring out what is the best course of action. Moreover, those actions must be well judged, whether in the lab, the newsroom, or the policy maker's office.

Of course it's true that many experts are decidedly inexpert at communicating their work in lay terms. This remains a major issue largely because experts are only very rarely called upon to engage in general-audience communication, hence they do not see gaining such skills as a priority. Training and advice are available, often from university press offices, but even when experts take advantage of such opportunities, they generally do so too little and too late.

However, in my view that's a secondary issue. As a scientist with the luxury of communicating frequently with the general public, I can report with confidence that experience helps only up to a point. A fundamental obstacle remains: Nonscientists harbor deep-seated instincts concerning the management of uncertainty in their everyday lives—instincts that exist because they generally work but that profoundly differ from the optimal strategy in science and technology. And of course it is technology that matters here, because technology is where the rubber hits the road—where science and the real world meet and must communicate effectively.

Examples of failure in this regard abound—so much so that they are hardly worthy of enumeration. Whether it be swine flu, bird flu, GM crops, or stem cells, the public debate departs so starkly from the scientist's comfort zone that it is hard not to sympathize with the errors scientists make, such as letting nuclear transfer be called "cloning," which end up holding critical research fields back for years.

One particular aspect of this problem stands out in its potential for public self-harm, however: risk aversion. When uncertainty revolves around such areas as ethics (as with nuclear transfer) or economic policy (as with flu vaccination), the issues are potentially avoidable by appropriate planning. This is not the case when it comes to the public attitude to risk. The immense decrease in vaccinations for major childhood diseases, following a single, controversial study linking them to autism, is a prime example. Another is the suspension of essentially all clinical trials of gene therapy for at least a year in response to the death of one person in a trial—a decision taken by regulatory bodies, yes, but one that was in line with public opinion.

These responses to the risk/benefit ratio of cutting-edge technologies are examples of fear of the unknown—of an irrationally conservative prioritization of the risks of change over the benefits, with unequivocally deleterious consequences in terms of quality and quantity of life in the future. Fear of the unknown is not remotely irrational in principle, when "fear of" is understood as a synonym for "caution about," but it can be and generally is overdone. If the public could be brought to a greater understanding of how to evaluate the risks inherent in exploring future technology, and the merits of accepting some short-term risk in the interests of overwhelmingly greater expected long-term benefit, progress in all areas of technology—especially biomedical technology—would be greatly accelerated.

BECAUSE

NIGEL GOLDENFELD
Professor of physics, University of Illinois–Urbana-Champaign

When you're facing in the wrong direction, progress means walking backward. History suggests that our worldview undergoes disruptive change not so much when science adds new concepts to our cognitive toolkit as when it takes away old ones. The sets of intuitions that have been with us since birth define our scientific prejudices, and they not only are poorly suited to the realms of the very large and very small but also fail to describe everyday phenomena. If we are to identify where the next transformation of our worldview will come from, we need to take a fresh look at our deep intuitions. In the two minutes it takes you to read this essay, I am going to try and rewire your basic thinking about causality.

Causality is usually understood as meaning that there is a single, preceding cause for an event. For example, in classical physics, a ball may be flying through the air because of having been hit by a tennis racket. My sixteen-year-old car always revs much too fast because the temperature sensor wrongly indicates that the engine temperature is cold, as if the car were in start-up mode. We are so familiar with causality as an underlying feature of reality that we hardwire it into the laws of physics. It might seem that this would be unnecessary, but it turns out that the laws of physics do not distinguish between time going backward and time going forward. And so we make a choice about which sort of physical law we would like to have.

However, complex systems, such as financial markets or the Earth's biosphere, do not seem to obey causality. For every event that occurs, there are a multitude of possible causes, and the extent to which each contributes to the event is not clear, not even after the fact! One might say that there is a web of causation. For example, on a typical day, the stock market might go up or down by some fraction of a percentage point. The *Wall Street Journal* might blithely report that the stock market move was due to "traders taking profits" or perhaps "bargain hunting by investors." The following day, the move might be in the opposite direction, and a different, perhaps contradictory, cause will be invoked. However, for each transaction there is both a buyer and a seller, and their worldviews must be opposite for the transaction to occur. Markets work only because there is a plurality of views. To assign a single or dominant cause to most market moves is to ignore the multitude of market outlooks and fail to recognize the nature and dynamics of the temporary imbalances between the numbers of traders who hold these differing views.

Similar misconceptions abound elsewhere in public debate and the sciences. For example, are there single causes for diseases? In some cases, such as Huntington's disease, the cause can be traced to a unique factor—in this case, extra repetitions of a particular nucleotide sequence at a particular location in an individual's DNA, coding for the amino acid glutamine. However, even in this case, the age of onset and the severity of the condition are also known to be controlled by environmental factors and interactions with other genes. The web of causation has been for many decades a well-worked metaphor in epidemiology, but there is still little quantitative understanding of how the web functions or forms. As Nancy Krieger of the Harvard School of Public Health poignantly asked in a celebrated 1994 essay, "Has anyone seen the spider?"

The search for causal structure is nowhere more futile than in the debate over the origin of organismal complexity: intelligent design vs. evolution. Fueling the debate is a fundamental notion of causality—that there is a beginning to life and that such a beginning must have had a single cause. On the other hand, if there is instead a web of causation driving the origin and evolution of life, a skeptic might ask: Has anyone seen the spider?

It turns out that there is no spider. Webs of causation can form spontaneously through the concatenation of associations between the agents or active elements in the system. For example, consider the Internet. Although a unified protocol for communication (TCP/IP, etc.) exists, the topology and structure of the Internet emerged during a frenzied build-out, as Internet service providers staked out territory in a gold rush of unprecedented scale. Remarkably, once the dust began to settle, it became apparent that the statistical properties of the resulting Internet were quite special: The time delays for packet transmission, the network topology, and even the information transmitted exhibit fractal properties.

However you look at the Internet—locally or globally, on short time scales or long—it looks exactly the same. Although the discovery of this fractal structure, around 1995, was an unwelcome surprise because standard traffic-control algorithms, as used by routers, were designed assuming that all properties of the network dynamics would be random, the fractality is also broadly characteristic of biological networks. Without a master blueprint, the evolution of an Internet is subject to the same underlying statistical laws that govern biological evolution, and structure emerges spontaneously, without the need for a controlling entity. Moreover, the resultant network can come to life in strange and unpredictable ways, obeying new laws whose origin cannot be traced to any one part of the network. The network behaves as a collective, not just the sum of parts, and to talk about causality is meaningless, because the behavior is distributed in space and in time.

Between 2:42 P.M. and 2:50 P.M. on May 6, 2010, the Dow-Jones Industrial Average experienced a rapid decline and subsequent rebound of nearly six hundred points, an event of unprecedented magnitude and brevity. This disruption occurred as part of a tumultuous event on that day now known as the Flash Crash, which affected numerous market indices and individual stocks, even causing some stocks to be priced at unbelievable levels (e.g., Accenture was at one point priced at $.01).

With tick-by-tick data available for every trade, we can watch the crash unfold in slow motion, a film of a financial calamity. But the cause of the crash itself remains a mystery. The U.S. Securities & Exchange Commission report on the Flash Crash was able to identify the trigger event (a $4 billion sale by a mutual fund) but could provide no detailed understanding of why this event caused the crash. The conditions that precipitated the crash were already embedded in the market's web of causation, a self-organized, rapidly evolving structure created by the interplay of high-frequency trading algorithms. The Flash Crash was the birth cry of a network coming to life, eerily reminiscent of Arthur C. Clarke's science fiction story "Dial F for Frankenstein," which begins "At 0150 GMT on December 1, 1975, every telephone in the world started to ring." I'm excited by the scientific challenge of understanding all this in detail, because . . . well, never mind. I guess I don't really know.

THE NAME GAME

STUART FIRESTEIN
Neuroscientist, chair of the Department of Biological Sciences, Columbia University

Too often in science we operate under the principle that "to name it is to tame it," or so we think. One of the easiest mistakes, even among working scientists, is to believe that labeling something has somehow or other added to an explanation or understanding of it. Worse than that, we use it all the time when we're teaching, leading students to believe that a phenomenon named is a phenomenon known, and that to know the name is to know the phenomenon. It's what I and others have called the nominal fallacy. In biology especially, we have labels for everything—molecules, anatomical parts, physiological functions, organisms, ideas, hypotheses. The nominal fallacy is the error of believing that the label carries explanatory information.

An instance of the nominal fallacy is most easily seen when the meaning or importance of a term or concept shrinks with knowledge. One example of this would be the word "instinct." "Instinct" refers to a set of behaviors whose actual cause we don't know, or simply don't understand or have access to, and therefore we call them instinctual, inborn, innate. Often this is the end of the exploration of these behaviors. They are the "nature" part of the nature-nurture argument (a term itself likely a product of the nominal fallacy) and therefore can't be broken down or reduced any further. But experience has shown that this is rarely the truth.

One of the great examples: It was for quite some time thought that when chickens hatched and immediately began pecking the ground for food, this behavior must have been instinctive. In the 1920s, a Chinese researcher named Zing-Yang Kuo made a remarkable set of observations on the developing chick egg that overturned this idea—and many similar ones. Using a technique of elegant simplicity, he found that rubbing heated Vaseline on a chicken egg caused it to become transparent enough so that he could see the embryo inside without disturbing it. In this way, he was able to make detailed observations of the chick's development, from fertilization to hatching. One of his observations was that in order for the growing embryo to fit properly in the egg, the neck is bent over the chest in such a way that the head rests on the chest just where the developing heart is encased. As the heart begins beating, the head of the chicken is moved up and down in a manner that precisely mimics the movement that will be used later for pecking the ground. Thus the "innate" pecking behavior that the chicken appears to know miraculously upon birth has, in fact, been practiced for more than a week within the egg.

In medicine, as well, physicians often use technical terms that lead patients to believe that more is known about pathology than may actually be the case. In Parkinson's patients, we note an altered gait and in general slower movements. Physicians call this bradykinesia, but it doesn't really tell you any more than if they simply said, "They move slower." Why do they move slower? What is the pathology and what is the mechanism for this slowed movement? These are the deeper questions hidden by the simple statement that "a cardinal symptom of Parkinson's is bradykinesia," satisfying though it may be to say the word to a patient's family.

In science, the one critical issue is to be able to distinguish between what we know and what we don't know. This is often difficult enough, as things that seem known sometimes become unknown—or at least more ambiguous. When is it time to quit doing an experiment because we now know something? When is it time to stop spending money and resources on a particular line of investigation because the facts are known? This line between the known and the unknown is already difficult enough to define, but the nominal fallacy often needlessly obscures it. Even words that, like "gravity," seem well settled may lend more of an aura to an idea than it deserves. After all, the apparently very-well-settled ideas of Newtonian gravity were almost completely undone after four hundred years by Einstein's general relativity. And still, today, physicists do not have a clear understanding of what gravity is or where it comes from, even though its effects can be described quite accurately.

Another facet of the nominal fallacy is the danger of using common words and giving them a scientific meaning. This has the often disastrous effect of leading an unwary public down a path of misunderstanding. Words like "theory," "law," and "force" do not mean in common discourse what they mean to a scientist. "Success" in Darwinian evolution is not the same "success" as taught by Dale Carnegie. "Force" to a physicist has a meaning quite different from that used in political discourse. The worst of these, though, may be "theory" and "law," which are almost polar opposites—theory being a strong idea in science while vague in common discourse, and law being a much more muscular social than scientific concept. These differences lead to sometimes serious misunderstandings between scientists and the public that supports their work.

Of course language is critical, and we must have names for things to talk about them. But the power of language to direct thought should never be taken lightly, and the dangers of the name game deserve our respect.

LIVING IS FATAL

SETH LLOYD
Quantum mechanical engineer, MIT; author, Programming the Universe

The ability to reason clearly in the face of uncertainty.

If everybody could learn to deal better with the unknown, this would improve not only their individual cognitive toolkit (to be placed in a slot right next to the ability to operate a remote control, perhaps) but the chances for humanity as a whole.

A well-developed scientific method for dealing with the unknown has existed for many years—the mathematical theory of probability. Probabilities are numbers whose values reflect how likely different events are to take place. People are bad at assessing probabilities. They are bad at it not just because they are bad at addition and multiplication. Rather, people are bad at probability on a deep, intuitive level: They overestimate the probability of rare but shocking events—a burglar breaking into your bedroom while you're asleep, say. Conversely, they underestimate the probability of common but quiet and insidious events—the slow accretion of globules of fat on the walls of an artery, or another ton of carbon dioxide pumped into the atmosphere.

I can't say I'm optimistic about the odds that people will learn to understand the science of odds. When it comes to understanding probability, people basically suck. Consider the following example, based on a true story and reported by Joel Cohen of Rockefeller University. A group of graduate students note that women have a significantly lower chance of admission than men to the graduate programs at a major university. The data are unambiguous: Women applicants are only two-thirds as likely as male applicants to be admitted. The graduate students file suit against the university, alleging discrimination on the basis of gender. When admissions data are examined on a department-by-department basis, however, a strange fact emerges: Within each department, women are *more* likely to be admitted than men. How can this possibly be?

The answer turns out to be simple, if counterintuitive. More women are applying to departments that have few positions. These departments admit only a small percentage of applicants, men or women. Men, by contrast, are applying to departments that have more positions and so admit a higher percentage of applicants. Within each department, women have a better chance of admission than men—it's just that few women apply to the departments that are easy to get into.

This counterintuitive result indicates that the admissions committees in the different departments are not discriminating against women. That doesn't mean that bias is absent. The number of graduate fellowships available in a particular field is determined largely by the federal government, which chooses how to allocate research funds to different fields. It is not the university that is guilty of sexual discrimination but the society as a whole, which chose to devote more resources—and so more graduate fellowships—to the fields preferred by men.

Of course, some people are good at probability. A car-insurance company that can't accurately determine the probabilities of accidents will go broke. In effect, when we pay premiums to insure ourselves against a rare event, we are buying into the insurance company's estimate of just how likely that event is. Driving a car, however, is one of those common but dangerous processes where human beings habitually underestimate the odds of something bad happening. Accordingly, some are disinclined to obtain car insurance (perhaps not suprising, when the considerable majority of people rate themselves as better-than-average drivers). When a state government requires its citizens to buy car insurance, it does so because it figures, rightly, that people are underestimating the odds of an accident.

Let's consider the debate over whether health insurance should be required by law. Living, like driving, is a common but dangerous process where people habitually underestimate risk, despite the fact that, with probability equal to 1, living is fatal.

UNCALCULATED RISK

GARRETT LISI
Independent theoretical physicist

We humans are terrible at dealing with probability. We are not merely bad at it but seem hardwired to be incompetent, in spite of the fact that we encounter innumerable circumstances every day which depend on accurate probabilistic calculations for our well-being. This incompetence is reflected in our language, in which the common words used to convey likelihood are "probably" and "usually"—vaguely implying a 50 to 100 percent chance. Going beyond the crude expression requires awkwardly geeky phrasing, such as "with 70 percent certainty," likely only to raise the eyebrow of a casual listener bemused by the unexpected precision. This blind spot in our collective consciousness—the inability to deal with probability—may seem insignificant, but it has dire practical consequences. We are afraid of the wrong things, and we are making bad decisions.

Imagine the typical emotional reaction to seeing a spider: fear, ranging from minor trepidation to terror. But what is the likelihood of dying from a spider bite? Fewer than four people a year (on average) die from spider bites, establishing the expected risk of death by spider at lower than 1 in 100 million. This risk is so minuscule that it is actually counterproductive to worry about it: Millions of people die each year from stress-related illnesses. The startling implication is that the risk of being bitten and killed by a spider is less than the risk that being afraid of spiders will kill you because of the increased stress.

Our irrational fears and inclinations are costly. The typical reaction to seeing a sugary doughnut is the desire to consume it.

But given the potential negative impact of that doughnut, including the increased risk of heart disease and reduction in overall health, our reaction should rationally be one of fear and revulsion. It may seem absurd to fear a doughnut—or, even more dangerous, a cigarette—but this reaction rationally reflects the potential negative effect on our lives.

We are especially ill-equipped to manage risk when dealing with small likelihoods of major events. This is evidenced by the success of lotteries and casinos at taking people's money, but there are many other examples. The likelihood of being killed by terrorism is extremely low, yet we have instituted actions to counter terrorism that significantly reduce our quality of life. As a recent example, X-ray body scanners could increase the risk of cancer to a degree greater than the risk from terrorism—the same sort of counterproductive overreaction as the one to spiders. This does not imply that we should let spiders, or terrorists, crawl all over us—but the risks need to be managed rationally.

Socially, the act of expressing uncertainty is a display of weakness. But our lives are awash in uncertainty, and rational consideration of contingencies and likelihoods is the only sound basis for good decisions. As another example, a federal judge recently issued an injunction blocking stem-cell research funding. The probability that stem-cell research will quickly lead to life-saving medicine is low, but if successful, the positive effects could be huge. If one considers outcomes and approximates the probabilities, the conclusion is that the judge's decision destroyed the lives of thousands of people, based on probabilistic expectation.

How do we make rational decisions based on contingencies? That judge didn't actually cause thousands of people to die . . . or did he? If we follow the "many worlds" interpretation of quantum physics—the most direct interpretation of its mathematical description—then our universe is continually branching into all possible contingencies: There is a world in which stem-cell research saves millions of lives and another world in which people die because of the judge's decision. Using the "frequentist" method of calculating probability, we have to add the probabilities of the worlds in which an event occurs to obtain the probability of that event.

Quantum mechanics dictates that the world we experience will happen according to this probability—the likelihood of the event. In this bizarre way, quantum mechanics reconciles the frequentist and "Bayesian" points of view, equating the frequency of an event over many possible worlds with its likelihood. An "expectation value," such as the expected number of people killed by the judge's decision, is the number of people killed in the various contingencies, weighted by their probabilities. This expected value is not necessarily likely to happen but is the weighted average of the expected outcomes—useful information when making decisions. In order to make good decisions about risk, we need to become better at these mental gymnastics, improve our language, and retrain our intuition.

Perhaps the best arena for honing our skills and making precise probabilistic assessments would be a betting market—an open site for betting on the outcomes of many quantifiable and socially significant events. In making good bets, all the tools and shorthand abstractions of Bayesian inference come into play—translating directly to the ability to make good decisions. With these skills, the risks we face in everyday life would become clearer and we would develop more rational intuitive responses to uncalculated risks, based on collective rational assessment and social conditioning.

We might get over our excessive fear of spiders and develop a healthy aversion to doughnuts, cigarettes, television, and stressful full-time employment. We would become more aware of the low cost, compared to probable rewards, of research, including research into improving the quality and duration of human life. And more subtly, as we became more aware and apprehensive of ubiquitous vague language such as "probably" and "usually," our standards of probabilistic description would improve.

Making good decisions requires concentrated mental effort, and if we overdo it, we run the risk of being counterproductive through increased stress and wasted time. So it's best to balance, and play, and take healthy risks—as the greatest risk is that we'll get to the end of our lives having never risked them on anything.

TRUTH IS A MODEL

NEIL GERSHENFELD

Physicist; director, MIT's Center for Bits and Atoms; author, Fab: The Coming Revolution on Your Desktop—From Personal Computers to Personal Fabrication

The most common misunderstanding about science is that scientists seek and find truth. They don't—they make and test models.

Kepler, packing Platonic solids to explain the observed motion of planets, made pretty good predictions, which were improved by his laws of planetary motion, which were improved by Newton's laws of motion, which were improved by Einstein's general relativity. Kepler didn't become wrong because of Newton's being right, just as Newton didn't then become wrong because of Einstein's being right; these successive models differed in their assumptions, accuracy, and applicability, not in their truth.

This is entirely unlike the polarizing battles that define so many areas of life: Either my political party, or religion, or lifestyle, is right or yours is, and I believe in mine. The only thing that's shared is the certainty of infallibility.

Building models is very different from proclaiming truths. It's a never-ending process of discovery and refinement, not a war to win or destination to reach. Uncertainty is intrinsic to the process of finding out what you don't know, not a weakness to avoid. Bugs are features—violations of expectations are opportunities to refine them. And decisions are made by evaluating what works better, not by invoking received wisdom.

These are familiar aspects of the work of any scientist, or baby: It's not possible to learn to talk or walk without babbling or toddling to experiment with language and balance. Babies who keep babbling turn into scientists who formulate and test theories for a living. But it doesn't require professional training to make mental models—we're born with those skills. What's needed is not displacing them with the certainty of absolute truths that inhibit the exploration of ideas. Making sense of anything means making models that can predict outcomes and accommodate observations. Truth is a model.

E PLURIBUS UNUM

JON KLEINBERG

Professor of computer science, Cornell University; coauthor (with David Easley), Networks, Crowds, and Markets: Reasoning About a Highly Connected World

If you used a personal computer twenty-five years ago, everything you needed to worry about was taking place in the box in front of you. Today, the applications you use over the course of an hour are scattered across computers all over the world; for the most part, we've lost the ability to tell where our data sit at all. We invent terms to express this lost sense of direction: Our messages, photos, and online profiles are all somewhere in "the cloud."

The cloud is not a single thing. What you think of as your Gmail account or Facebook profile is in fact made possible by the teamwork of a huge number of physically dispersed components—a distributed system, in the language of computer science. But we can think of it as a single thing, and this is the broader point: The ideas of distributed systems apply whenever we see many small things working independently but cooperatively to produce the illusion of a single unified experience. This effect takes place not just on the Internet but in many other domains as well. Consider, for example, a large corporation that releases new products and makes public announcements as though it were a single actor, when we know that at a more detailed level it consists of tens of thousands of employees. Or a massive ant colony engaged in coordinated exploration, or the neurons of your brain creating your experience of the present moment.

The challenge for a distributed system is to achieve this illusion of a single unified behavior in the face of so much underlying complexity. And this broad challenge, appropriately, is in fact composed of many smaller challenges in tension with one another.

One basic piece of the puzzle is the problem of consistency. Each component of a distributed system sees different things and has a limited ability to communicate with everything else, so different parts of the system can develop views of the world that are mutually inconsistent. There are numerous examples of how this can lead to trouble, both in technological domains and beyond. Your handheld device doesn't sync with your e-mail, so you act without realizing that there's already been a reply to your message. Two people across the country both reserve seat 5F on the same flight at the same time. An executive in an organization "didn't get the memo" and so strays off-message. A platoon attacks too soon and alerts the enemy.

It is natural to try "fixing" these kinds of problems by enforcing a single global view of the world and requiring all parts of the system to constantly refer to this global view before acting. But this undercuts many of the reasons you use a distributed system in the first place. It makes the component that provides the global view an enormous bottleneck and a highly dangerous single point of potential failure. The corporation doesn't function if the CEO has to sign off on every decision.

To get a more concrete sense of some of the underlying design issues, it helps to walk through an example in a little detail—a basic kind of situation, in which we try to achieve a desired outcome with information and actions that are divided among multiple participants. The example is the problem of sharing information securely: Imagine trying to back up a sensitive database on multiple computers while protecting the data so that it can be reconstructed only if a majority of the backup computers cooperate. But since the question of secure information-sharing ultimately has nothing specifically to do with computers or the Internet, let's formulate it instead using a story about a band of pirates and a buried treasure.

Suppose that an aging pirate king knows the location of a secret treasure and before retiring he intends to share the secret among his five shiftless sons. He wants them to be able to recover the treasure if three or more of them work together, but he also wants to prevent a "splinter group" of one or two from being able to get the treasure on their own. To do this, he plans to split the secret of the location into five "shares," giving one to each son, in such a way that he ensures the following condition. If, at any point in the future, at least three of the sons pool their shares of the secret, then they will know enough to recover the treasure. But if only one or two pool their shares, they will not have enough information.

How to do this? It's not hard to invent ways of creating five clues so that all of them are necessary for finding the treasure. But this would require unanimity among the five sons before the treasure could be found. How can we do it so that cooperation among any three is enough and cooperation among any two is insufficient?

Like many deep insights, the answer is easy to understand in retrospect. The pirate king draws a secret circle on the globe (known only to himself) and tells his sons that he's buried the treasure at the exact southernmost point on this circle. He then gives each son a different point on this circle. Three points are enough to uniquely reconstruct a circle, so any three pirates can pool their information, identify the circle, and find the treasure. But for any two pirates, an infinity of circles pass through their two points, and they cannot know which is the one they need for recovering the secret. It's a powerful trick and broadly applicable; in fact, versions of this secret-sharing scheme form a basic principle of modern data security, discovered by the cryptographer Adi Shamir, wherein arbitrary data can be encoded using points on a curve and reconstructed from knowledge of other points on the same curve.

The literature on distributed systems is rich with ideas in this spirit. More generally, the principles of distributed systems give us a way to reason about the difficulties inherent in complex systems built from many interacting parts. And so to the extent that we sometimes are fortunate enough to get the impression of a unified Web, a unified global banking system, or a unified sensory experience, we should think about the myriad challenges involved in keeping these experiences whole.

A PROXEMICS OF URBAN SEXUALITY

STEFANO BOERI

Architect, Politecnico of Milan; visiting professor, Harvard University Graduate School of Design; editor-in-chief, Abitare *magazine*

In every room, in every house, in every street, in every city, movements, relations, and spaces are also defined with regard to logics of sexual attraction-repulsion between individuals. Even the most insurmountable ethnic or religious barriers can suddenly disappear in the furor of intercourse; even the warmest and most cohesive community can rapidly dissolve in the absence of erotic tension. To understand how our cosmopolitan and multigendered cities work, we need a proxemics of urban sexuality.

FAILURE LIBERATES SUCCESS

KEVIN KELLY
Editor-at-large, Wired *magazine; author,* What Technology Wants

We can learn nearly as much from an experiment that doesn't work as from one that does. Failure is not something to be avoided but something to be cultivated. That's a lesson from science that benefits not only laboratory research but design, sport, engineering, art, entrepreneurship, and even daily life itself. All creative avenues yield the maximum when failures are embraced. A great graphic designer will generate lots of ideas, knowing that most will be aborted. A great dancer realizes that most new moves will not succeed. Ditto for any architect, electrical engineer, sculptor, marathoner, startup maven, or microbiologist. What is science, after all, but a way to learn from things that don't work, rather than just those that do? What this tool suggests is that you should aim for success while being prepared to learn from a series of failures. More so, you should carefully but deliberately press your successful investigations or accomplishments to the point where they break, flop, stall, crash, or fail.

Failure was not always so noble. In fact, in much of the world today, failure is still not embraced as a virtue. It is a sign of weakness and often a stigma that prohibits second chances. Children in many parts of the world are taught that failure brings disgrace and that one should do everything in one's power to succeed without failure. Yet the rise of the West is in many respects due to the rise in tolerating failure. Indeed, many immigrants trained in a failure-intolerant culture may blossom out of stagnancy once moved into a failure-tolerant culture. Failure liberates success.

The chief innovation that science brought to the state of defeat is a way to manage mishaps. Blunders are kept small, manageable, constant, and trackable. Flops are not quite deliberate, but they are channeled so that something is learned each time things fail. It becomes a matter of failing forward. Science itself is learning how to better exploit negative results. Due to the problems of costly distribution, most negative results have not been shared, thus limiting their potential to speed learning for others. But increasingly published negative results (which include experiments that succeed in showing no effects) are becoming another essential tool in the scientific method.

Wrapped up in the idea of embracing failure is the related notion of breaking things to make them better—particularly complex things. Often the only way to improve a complex system is to probe its limits by forcing it to fail in various ways. Software, among the most complex things we make, is usually tested for quality by employing engineers to systematically find ways to crash it. Similarly, one way to troubleshoot a complicated device that's broken is to deliberately force negative results (temporary breaks) in its multiple functions in order to locate the actual dysfunction. Great engineers have a respect for breaking things that sometimes surprises nonengineers, just as scientists have a patience with failures that often perplexes outsiders. But the habit of embracing negative results is one of the most essential tricks to gaining success.

HOLISM

NICHOLAS A. CHRISTAKIS
Physician and social scientist, Harvard University; coauthor (with James H. Fowler), Connected: The Surprising Power of Our Social Networks and How They Shape Our Lives

Some people like to build sand castles and some like to tear them apart. There can be much joy in the latter, but it is the former that interests me. You can take a bunch of minute silica crystals, pounded for thousands of years by the waves, use your hands, and make an ornate tower. Tiny physical forces govern how each particle interacts with its neighbors, keeping the castle together—at least until the force majeure of a foot appears. But this is the part I like most: Having built the castle, you step back and look at it. Across the expanse of beach, here is something new, something not present before among the endless sand grains, something risen from the ground, something that reflects the scientific principle of holism.

Holism is colloquially summarized as "The whole is greater than the sum of its parts." What interests me, however, are not the artificial instantiations of this principle—when we deliberately form sand into ornate castles, or metal into airplanes, or ourselves into corporations—but rather the natural instantiations. Examples are widespread and stunning. Perhaps the most impressive is that carbon, hydrogen, oxygen, nitrogen, sulfur, phosphorus, iron, and a few other elements, mixed in just the right way, yield life. And life has emergent properties not present in or predictable from these constituent parts. There is a kind of awesome synergy between the parts.

Hence, I think that the scientific concept that would improve everybody's cognitive toolkit is holism: the abiding recognition that wholes have properties not present in the parts and not reducible to the study of the parts.

For example, carbon atoms have particular, knowable physical and chemical properties. But the atoms can be combined in different ways to make, say, graphite or diamonds. The properties of those substances—properties such as darkness and softness and clearness and hardness—are properties not of the carbon atoms but rather of the collection of carbon atoms. Moreover, which particular properties the collection of atoms has depends entirely on how they are assembled—into sheets or pyramids. The properties arise because of the connections between the parts. Grasping this insight is crucial for a proper scientific perspective on the world. You could know everything about isolated neurons and be unable to say how memory works or where desire originates.

It is also the case that the whole has a complexity that rises faster than the number of its parts. Consider social networks as a simple illustration. If we have 10 people in a group, there are a maximum of 10 x 9/2 = 45 possible connections between them. If we increase the number of people to 1,000, the number of possible ties increases to 1,000 x 999/2 = 499,500. So, while the number of people has increased by a hundredfold (from 10 to 1,000), the number of possible ties (and hence this one measure of the system's complexity) has increased more than ten thousandfold.

Holism does not come naturally. It is an appreciation not of the simple but of the complex—or, at least, of the simplicity and coherence in complex things. Unlike curiosity or empiricism, say, holism takes a while to acquire and appreciate. It is a grown-up disposition. Indeed, for the last few centuries the Cartesian project in science has been to break matter down into ever smaller bits in the pursuit of understanding. And this works to some extent. We can understand matter by breaking it down to atoms, then protons and electrons and neutrons, then quarks, then gluons, and so on. We can understand organisms by breaking them down into organs, then tissues, then cells, then organelles, then proteins, then DNA, and so on.

Putting things back together in order to understand them is harder and typically comes later in the development of a scientist or of science. Think of the difficulties in understanding how all the cells in our bodies work together, as compared with the study of the cells themselves. Whole new fields of neuroscience and systems biology and network science are arising to accomplish just this. And these fields are arising just now, after centuries of stomping on castles in order to figure them out.

TANSTAAFL

ROBERT R. PROVINE

Psychologist and neuroscientist, University of Maryland; author, Laughter: A Scientific Investigation

TANSTAAFL is the acronym for "There ain't no such thing as a free lunch," a universal truth having broad and deep explanatory power in science and daily life. The expression originated from the practice of saloons offering free lunch if you bought their overpriced drinks. Science fiction master Robert Heinlein introduced me to TANSTAAFL in *The Moon Is a Harsh Mistress*, his 1966 classic in which a character warns of the hidden cost of a free lunch.

The universality of the fact that you can't get something for nothing has found application in sciences as diverse as physics (the laws of thermodynamics) and economics, where Milton Friedman used a grammatically upgraded variant as the title of his 1975 book *There's No Such Thing as a Free Lunch*. Physicists are clearly on board with TANSTAAFL; less so, many political economists in their smoke-and-mirrors world.

My students hear a lot about TANSTAAFL—from the biological costs of the peacock's tail to our nervous system, which distorts physical reality to emphasize changes in time and space. When the final tally is made, peahens cast their ballot for the sexually exquisite plumage of the peacock and its associated vigor; likewise, it is more adaptive for humans to detect critical sensory events than to be high-fidelity light and sound meters. In such cases, lunch comes at reasonable cost, as determined by the grim but honest accounting of natural selection, a process without hand waving and incantation.

SKEPTICAL EMPIRICISM

GERALD HOLTON

Mallinckrodt Professor of Physics and professor of the history of science, emeritus, Harvard University; coeditor, Einstein for the 21st Century: His Legacy in Science, Art and Modern Culture

In politics and society at large, important decisions are all too often based on deeply held presuppositions, ideology, or dogma—or, on the other hand, on headlong pragmatism without study of long-range consequences.

Therefore I suggest the adoption of *skeptical empiricism*, the kind exemplified by the carefully thought-out and tested research in science at its best. It differs from plain empiricism of the sort that characterized the writings of the scientist/philosopher Ernst Mach, who refused to believe in the existence of atoms because one could not "see" them.

To be sure, in politics and daily life, on some topics decisions have to be made very rapidly on few or conflicting data. Yet precisely for that reason, it will be wise also to launch a more considerate program of skeptical empiricism on the same topic, if only to be better prepared for the consequences, intended or not, that followed from the quick decision.

OPEN SYSTEMS

THOMAS A. BASS
Professor of English, State University of New York–Albany; author, The Spy Who Loved Us

This year, *Edge* is asking us to identify a scientific concept that "would improve everybody's cognitive toolkit." Not clever enough to invent a concept of my own, I am voting for a winning candidate. It might be called the Swiss Army knife of scientific concepts, a term containing a remarkable number of useful tools for exploring cognitive conundrums. I am thinking of open systems, an idea that passes through thermodynamics and physics before heading into anthropology, linguistics, history, philosophy, and sociology, until arriving, finally, into the world of computers, where it branches into other ideas, such as open source and open standards.

Open standards allow knowledgeable outsiders access to the design of computer systems, to improve, interact with, or otherwise extend them. These standards are public, transparent, widely accessible, and royalty-free for developers and users. Open standards have driven innovation on the Web and allowed it to flourish as both a creative and commercial space.

Unfortunately, the ideal of an open Web is not embraced by companies that prefer walled gardens, silos, proprietary systems, apps, tiered levels of access, and other metered methods for turning citizens into consumers. Their happy-face Web contains tracking systems useful for making money, but these systems are also appreciated by the police states of the world, for they, too, have a vested interest in surveillance and closed systems.

Now that the Web has frothed through twenty years of chaotic inventiveness, we have to push back against the forces that would close it down. A similar push should be applied to other systems veering toward closure. "Citoyens, citoyennes, arm yourselves with the concept of openness."

NON-INHERENT INHERITANCE

GEORGE CHURCH
Professor, Harvard University; director, Personal Genome Project

The names Lysenko and Lamarck are nearly synonymous with bad science—worse than merely mediocre science because of the huge political and economical consequences.

From 1927 to 1964, Trofim Lysenko managed to keep the theory of the inheritance of acquired characteristics while dogmatically directing Soviet agriculture and science. Andrei Sakharov and other Soviet physicists finally provoked the fall of this cabal in the 1960s, blaming it for the "shameful backwardness of Soviet biology and of genetics in particular . . . defamation, firing, arrest, even death, of many genuine scientists."

At the opposite (yet equally discredited) end of the genetic-theory spectrum was the Galtonian eugenics movement, which from 1883 onward grew in popularity in many countries until the 1948 Universal Declaration of Human Rights ("the most translated document in the world") stated that "Men and women of full age, without any limitation due to race, nationality or religion, have the right to marry and to found a family." Nevertheless, forced sterilizations persisted into the 1970s. The "shorthand abstraction" is that Lysenkoism overestimated the impact of environment and eugenics overestimated the role of genetics.

One form of scientific blindness occurs, as above, when a theory displays exceptional political or religious appeal. But another source of blindness arises when we rebound from catastrophic failures of pseudoscience (or science). We might conclude from the two aforementioned genetic disasters that we need only to police abuses of our human germ line inheritance. Combining the above with the ever-simmering debate on Darwin, we might develop a bias that human evolution has stopped or that "design" has no role. But we are well into an unprecedented new phase of evolution, in which we must generalize beyond our DNA-centric worldview. We now inherit acquired characteristics. We always have, but now this feature is dominant and exponential. We apply eugenics at the individual family level (where it is a right), not the governmental level (where it is a wrong). Moreover, we might aim for the same misguided targets that eugenics chose (i.e., uniformity around "ideal" traits), via training and medications.

Evolution has accelerated from geologic speed to Internet speed—still employing random mutation and selection but also using nonrandom intelligent design, which makes it even faster. We are losing species not just by extinction but by merger. There are no longer species barriers between humans, bacteria, and plants—or even between humans and machines.

Shorthand abstractions are only one device we employ to construct the "Flynn effect"—the worldwide increase in average scores on intelligence tests. How many of us noticed the minor milestone when the SAT tests first permitted calculators? How many of us have participated in conversations semi-discreetly augmented by Google or text messaging? Even without invoking artificial intelligence, how far are we from commonplace augmentation of our decision making, the way we have augmented our math, memory, and muscles?

SHIFTING BASELINE SYNDROME

PAUL KEDROSKY
Editor, Infectious Greed; *senior fellow, Kauffman Foundation*

When John Cabot came to the Grand Banks off Newfoundland in 1497, he was astonished at what he saw. Fish, so many fish—fish in numbers he could hardly comprehend. According to Farley Mowat, Cabot wrote that the waters were so "swarming with fish [that they] could be taken not only with a net but in baskets let down and [weighted] with a stone." The fisheries boomed for five hundred years, but by 1992 it was all over. The Grand Banks cod fishery was destroyed, and the Canadian government was forced to close it entirely, putting thirty thousand fishers out of work. It has never recovered.

What went wrong? Many things, from factory fishing to inadequate oversight, but much of it was aided and abetted by treating each step toward disaster as normal. The entire path, from plenitude to collapse, was taken as the status quo, right up until the fishery was essentially wiped out.

In 1995, fisheries scientist Daniel Pauly coined a phrase for this troubling ecological obliviousness; he called it "shifting baseline syndrome." Here is how Pauly first described the syndrome:

> [E]ach generation of fisheries scientists accepts as a baseline the stock size and species composition that occurred at the beginning of their careers, and uses this to evaluate changes. When the next generation starts its career, the stocks have further declined, but it is the stocks at that time that serve as a new baseline. The result obviously is a gradual shift of the baseline, a gradual accommodation of the creeping disappearance of resource species.

It is blindness, stupidity, intergeneration data obliviousness. Most scientific disciplines have long timelines of data, but many ecological disciplines don't. We are forced to rely on secondhand and anecdotal information. We don't have enough data to know what is normal, so we convince ourselves that this is normal.

But it often isn't normal. Instead, it is a steadily and insidiously shifting baseline, no different from convincing ourselves that winters have always been this warm or this snowy. Or convincing ourselves that there have always been this many deer in the forests of eastern North America. Or that current levels of energy consumption per capita in the developed world are normal. All of these are shifting baselines, where our data inadequacy, whether personal or scientific, provides dangerous cover for missing important longer-term changes in the world around us.

When you understand shifting baseline syndrome, it forces you to continually ask what is normal. Is this? Was that? And, at least as important, it asks how we "know" that it's normal. Because if it isn't, we need to stop shifting the baselines and do something about it before it's too late.

PERMA

MARTIN SELIGMAN
Zellerbach Family Professor of Psychology and director of the Positive Psychology Center, University of Pennsylvania; author, Flourish: A Visionary New Understanding of Happiness and Well-Being

Is global well-being possible?

Scientists commonly predict dystopias: nuclear war, overpopulation, energy shortage, dysgenic selection, widespread sound-bite mentality, and the like. You don't get much attention predicting that the human future will work out. I am not, however, going to predict that a positive human future will in fact occur, but it becomes more likely if we think about it systematically. We can begin by laying out the measurable elements of well-being and then asking how those elements might be achieved. I address only measurement.

Well-being is about what individuals and societies choose for its own sake, that which is north of indifference. The elements of well-being must be exclusive, measurable independently of one another, and—ideally—exhaustive. I believe there are five such elements, and they have a handy acronym, PERMA:

P Positive Emotion
E Engagement
R Positive Relationships
M Meaning and Purpose
A Accomplishment

There has been forward movement in the measurement of these over the last decade. Taken together, PERMA forms a more comprehensive index of well-being than "life satisfaction," and it allows for the combining of objective and subjective indicators.

PERMA can index the well-being of individuals, of corporations, of cities. The United Kingdom has now undertaken the measurement of well-being for the nation and as one criterion—in addition to gross domestic product—of the success of its public policy.

PERMA is a shorthand abstraction for the enabling conditions of life.

How do the *disabling* conditions—such as poverty, disease, depression, aggression, and ignorance—relate to PERMA? The disabling conditions of life obstruct PERMA, but they do not obviate it. Importantly, the correlation of depression to happiness is not minus 1.00, it is only about minus 0.35, and the effect of income on life satisfaction is markedly curvilinear, with increasing income producing less and less life satisfaction the further above the safety net you are.

Science and public policy have traditionally been focused solely on remediating the disabling conditions, but PERMA suggests that this is insufficient. If we want global well-being, we should also measure, and try to build, PERMA. The very same principle seems to be true in your own life: If you wish to flourish personally, getting rid of depression, anxiety, and anger and amassing riches is not enough—you also need to build PERMA directly.

What is known about how PERMA can be built?

Perhaps the *Edge* Question for 2012 will be "How can science contribute to building global well-being?"

POSITIVE-SUM GAMES

STEVEN PINKER

Johnstone Family Professor, Department of Psychology, Harvard University; author, The Stuff of Thought: Language as a Window into Human Nature

A zero-sum game is an interaction in which one party's gain equals the other party's loss—the sum of their gains and losses is zero. (More accurately, it is constant across all combinations of their courses of action.) Sports matches are quintessential examples of zero-sum games: Winning isn't everything, it's the only thing, and nice guys finish last. A nonzero-sum game is an interaction in which some combinations of actions provide a net gain (positive sum) or loss (negative sum) to the two participants. The trading of surpluses, as when herders and farmers exchange wool and milk for grain and fruit, is a quintessential example, as is the trading of favors, as when people take turns baby-sitting each other's children.

In a zero-sum game, a rational actor seeking the greatest gain for himself or herself will necessarily be seeking the maximum loss for the other actor. In a positive-sum game, a rational, self-interested actor may benefit the other actor with the same choice that benefits himself or herself. More colloquially, positive-sum games are called win-win situations and are captured in the cliché "Everybody wins."

This family of concepts—zero-sum, nonzero-sum, positive-sum, negative-sum, constant-sum, and variable-sum games—was introduced by John von Neumann and Oskar Morgenstern when they invented the mathematical theory of games in 1944. The Google Books Ngram tool shows that the terms saw a steady increase in popularity beginning in the 1950s, and their colloquial relative "win-win" began a similar ascent in the 1970s.

Once people are thrown together in an interaction, their choices don't determine whether they are in a zero- or nonzero-sum game; the game is a part of the world they live in. But by neglecting some of the options on the table, people may perceive that they are in a zero-sum game when in fact they are in a nonzero-sum game. Moreover, they can change the world to make their interaction nonzero-sum. For these reasons, when people become aware of the game-theoretic structure of their interaction (that is, whether it is positive-, negative-, or zero-sum), they can make choices that bring them valuable outcomes—like safety, harmony, or prosperity—without their having to become more virtuous or noble.

Some examples: Squabbling colleagues or relatives agree to swallow their pride, take their losses, or lump it to enjoy the resulting comity rather than absorbing the costs of continual bickering in hopes of prevailing in a battle of wills. Two parties in a negotiation split the difference in their initial bargaining positions to "get to yes." Divorcing spouses realize they can reframe their negotiations: from each trying to get the better of the other while enriching the lawyers to trying to keep as much money as possible for the two of them and out of the billable hours of Dewey, Cheatham & Howe. Populaces recognize that economic middlemen (particularly ethnic minorities who specialize in that niche, such as Jews, Armenians, overseas Chinese, and expatriate Indians) are not social parasites whose prosperity comes at the expense of their hosts but positive-sum-game creators who enrich everyone at once. Countries recognize that international trade doesn't benefit their trading partner to their own detriment but benefits them both and turn away from beggar-thy-neighbor protectionism to open economies which (as classical economists noted) make everyone richer and (as political scientists have recently shown) discourage war and genocide. Warring countries lay down their arms and split the peace dividend rather than pursuing Pyrrhic victories.

Granted, some human interactions really are zero-sum; competition for mates is a biologically salient example. And even in positive-sum games, a party may pursue an individual advantage at the expense of joint welfare. But a full realization of the risks and costs of the game-theoretic structure of an interaction (particularly if it is repeated, so that the temptation to pursue an advantage in one round may be penalized when roles reverse in the next) can militate against various forms of short-sighted exploitation.

Has an increasing awareness of the zero- or nonzero-sumness of interactions in the decades since 1950 (whether referred to in those terms or not) actually led to increased peace and prosperity in the world? It's not implausible. International trade and membership in international organizations has soared in the decades that game-theoretic thinking has infiltrated popular discourse. And perhaps not coincidentally, the developed world has seen both spectacular economic growth and a historically unprecedented decline in several forms of institutionalized violence, such as war between great powers, war between wealthy states, genocides, and deadly ethnic riots. Since the 1990s, these gifts have started to accrue in the developing world as well, in part because those countries have switched their foundational ideologies from ones that glorify zero-sum class and national struggle to ones that glorify positive-sum market cooperation. (All these claims can be documented from the literature in international studies.)

The enriching and pacifying effects of participation in positive-sum games long antedate the contemporary awareness of the concept. The biologists John Maynard Smith and Eörs Szathmáry have argued that an evolutionary dynamic that creates positive-sum games drove the major transitions in the history of life: the emergence of genes, chromosomes, bacteria, cells with nuclei, organisms, sexual reproduction, and animal societies. In each transition, biological agents entered into larger wholes in which they specialized, exchanged benefits, and developed safeguards to prevent one from exploiting the rest to the detriment of the whole. The journalist Robert Wright has sketched a similar arc in his book *Nonzero* and extended it to the deep history of human societies. An explicit recognition among literate people of the shorthand abstraction "positive-sum game" and its relatives may be extending a process in the world of human choices that has been operating in the natural world for billions of years.

THE SNUGGLE FOR EXISTENCE

ROGER HIGHFIELD
Editor, New Scientist; *coauthor (with Martin Nowak),* SuperCooperators: Altruism, Evolution, and Why We Need Each Other to Succeed

Everyone is familiar with the struggle for existence. In the wake of the revolutionary work by Charles Darwin, we realized that competition is at the very heart of evolution. The fittest win this endless "struggle for life most severe," as he put it, and all others perish. In consequence, every creature that crawls, swims, and flies today has ancestors that once successfully reproduced more often than their unfortunate competitors.

This is echoed in the way that people see life as competitive. Winners take all. Nice guys finish last. We look after number one. We are motivated by self-interest. Indeed, even our genes are said to be selfish.

Yet competition does not tell the whole story of biology.

I doubt many realize that, paradoxically, one way to win the struggle for existence is to pursue the snuggle for existence: to cooperate.

We already do this to a remarkable extent. Even the simplest activities of everyday life involve much more cooperation than you might think. Consider, for example, stopping at a coffee shop one morning to have a cappuccino and croissant for breakfast. To enjoy that simple pleasure could draw on the labors of a small army of people from at least half a dozen countries. Delivering that snack also relied on a vast number of ideas, which have been widely disseminated around the world down the generations by the medium of language.

Now we have remarkable new insights into what makes us all work together. Building on the work of many others, Martin Nowak of Harvard University has identified at least five basic mechanisms of cooperation. What I find stunning is that he shows that the way we human beings collaborate is as clearly described by mathematics as the descent of the apple that once fell in Newton's garden. The implications of this new understanding are profound.

Global human cooperation now teeters on a threshold. The accelerating wealth and industry of Earth's increasing inhabitants—itself a triumph of cooperation—is exhausting the ability of our home planet to support us all. Many problems that challenge us today can be traced back to a profound tension between what is good and desirable for society as a whole and what is good and desirable for an individual. That conflict can be found in global problems such as climate change, pollution, resource depletion, poverty, hunger, and overpopulation.

As once argued by the American ecologist Garrett Hardin, the biggest issues of all—saving the planet and maximizing the collective lifetime of the species *Homo sapiens*—cannot be solved by technology alone. If we are to win the struggle for existence and avoid a precipitous fall, there's no choice but to harness this extraordinary creative force. It is down to all of us to refine and extend our ability to cooperate.

Nowak's work contains a deeper message. Previously there were only two basic principles of evolution—mutation and selection—where the former generates genetic diversity and the latter picks the individuals best suited to a given environment. We must now accept that cooperation is the third principle. From cooperation can emerge the constructive side of evolution, from genes to organisms to language and the extraordinarily complex social behaviors that underpin modern society.

THE LAW OF COMPARATIVE ADVANTAGE

DYLAN EVANS
Lecturer in behavioral science, School of Medicine, University College Cork, Ireland; author, Introducing Evolutionary Psychology: A Graphic Guide

It is not hard to identify the discipline in which to look for the scientific concept that would most improve everybody's cognitive toolkit; it *has* to be economics. No other field of study contains so many ideas ignored by so many people at such great cost to themselves and the world. The hard task is picking just one of the many such ideas that economists have developed.

On reflection, I have plumped for the law of comparative advantage, which explains how trade can be beneficial for both parties even when one of them is more productive than the other in every way. At a time of growing protectionism, it is more important than ever to reassert the value of free trade. Since trade in labor is roughly the same as trade in goods, the law of comparative advantage also explains why immigration is almost always a good thing—a point which also needs emphasizing at a time when xenophobia is on the rise.

In the face of well-meaning but ultimately misguided opposition to globalization, we must celebrate the remarkable benefits which international trade has brought us and fight for a more integrated world.

STRUCTURED SERENDIPITY

JASON ZWEIG
Journalist; personal finance columnist, Wall Street Journal; *author,* Your Money and Your Brain

Creativity is a fragile flower, but perhaps it can be fertilized with systematic doses of serendipity. The psychologist Sarnoff Mednick showed decades ago that some people are better than others at detecting the associations that connect seemingly random concepts. Asked to name a fourth idea that links "wheel," "electric," and "high," people who score well on other measures of creativity will promptly answer "chair." More recently, research in Mark Jung-Beeman's cognitive neuroscience lab at Northwestern has found that sudden bursts of insight—the *Aha!* or *Eureka!* moment—come when brain activity abruptly shifts its focus. The almost ecstatic sense that makes us cry "I see!" appears to come when the brain is able to shunt aside immediate or familiar visual inputs.

That may explain why so many of us close our eyes (often unwittingly) just before we exclaim, "I see!" It also suggests, at least to me, that creativity can be enhanced deliberately through environmental variation. Two techniques seem promising: varying what you learn and varying where you learn it. I try each week to read a scientific paper in a field new to me—and to read it in a different place.

New associations often leap out of the air at me this way. More intriguing, others seem to form covertly and lie in wait for the opportune moment when they can click into place. I do not try to force these associations out into the open; they are like the shrinking mimosa plants that crumple if you touch them but bloom if you leave them alone.

The sociologist Robert Merton argued that many of the greatest discoveries of science have sprung from serendipity. As a layman and an amateur, all I hope to accomplish by throwing myself in serendipity's path is to pick up new ideas, and combine old ones, in ways that haven't quite occurred to other people yet. So I let my curiosity lead me wherever it seems to want to go, like the planchette that floats across the Ouija board.

I do this remote-reading exercise on my own time, since it would be hard to justify to newspaper editors during the work day. But my happiest moments last year came as I reported an investigative article on how elderly investors are increasingly being scammed by elderly con artists. I later realized, to my secret delight, that the article had been enriched by a series of papers I had been reading on altruistic behavior among fish (*Lambroides dimidiatus*).

If I do my job right, my regular readers will never realize that I spend a fair amount of my leisure time reading *Current Biology*, *The Journal of Neuroscience*, and *Organizational Behavior and Human Decision Processes*. If that reading helps me find new ways to understand the financial world, as I suspect it does, my readers will indirectly be smarter for it. If not, the only harm done is my own spare time wasted.

In my view, we should each invest a few hours a week in reading research that ostensibly has nothing to do with our day jobs, in a setting that has nothing in common with our regular workspaces. This kind of structured serendipity just might help us become more creative, and I doubt that it can hurt.

THE WORLD IS UNPREDICTABLE

RUDY RUCKER

Mathematician; computer scientist; cyberpunk pioneer; novelist; author, Jim and the Flims

The media cast about for the proximate causes of life's windfalls and disasters. The public demands blocks against the bad and pipelines to the good. Legislators propose new regulations, fruitlessly dousing last year's fires, forever betting on yesterday's winning horses.

A little-known truth: Every aspect of the world is fundamentally unpredictable. Computer scientists have long since proved this.

How so? To predict an event is to know a shortcut for foreseeing the outcome. A simple counting argument shows that there aren't enough shortcuts to go around. Therefore most processes aren't predictable. A deeper argument plays on the fact that if you could predict your actions, you could deliberately violate your predictions, which means the predictions were wrong after all.

We often suppose that unpredictability is caused by random inputs from higher spirits or from lowdown quantum foam. But chaos theory and computer science tell us that nonrandom systems produce surprises on their own. The unexpected tornado, the cartoon safe that lands on Uncle George, the winning pull on a slot machine—odd things pop out of a computation. The world can simultaneously be deterministic and unpredictable.

In the physical world, the only way to learn tomorrow's weather in detail is to wait twenty-four hours and see, even if nothing is random at all. The universe is computing tomorrow's weather as rapidly and as efficiently as possible; any smaller model is inaccurate, and the smallest error is amplified into large effects.

At a personal level, even if the world is as deterministic as a computer program, you still can't predict what you're going to do. This is because your prediction method would involve a mental simulation of you that produces its results slower than you do. You can't think faster than you think. You can't stand on your own shoulders.

It's a waste to chase the pipe dream of a magical tiny theory that allows us to make quick and detailed calculations about the future. We can't predict and we can't control. To accept this can be a source of liberation and inner peace. We're part of the unfolding world, surfing the chaotic waves.

RANDOMNESS

CHARLES SEIFE
Professor of journalism, New York University; former journalist, Science; *author,* Proofiness: The Dark Arts of Mathematical Deception

Our very brains revolt at the idea of randomness. We have evolved as a species to become exquisite pattern-finders; long before the advent of science, we figured out that a salmon-colored sky heralds a dangerous storm or that a baby's flushed face likely means a difficult night ahead. Our minds automatically try to place data in a framework that allows us to make sense of our observations and use them to understand and predict events.

Randomness is so difficult to grasp because it works against our pattern-finding instincts. It tells us that sometimes there is no pattern to be found. As a result, randomness is a fundamental limit to our intuition; it says that there are processes we can't predict fully. It's a concept that we have a hard time accepting, even though it's an essential part of the way the cosmos works. Without an understanding of randomness, we are stuck in a perfectly predictable universe that simply doesn't exist outside our heads.

I would argue that only once we understand three dicta—three laws of randomness—can we break out of our primitive insistence on predictability and appreciate the universe for what it is, rather than what we want it to be.

The First Law of Randomness: There is such a thing as randomness.
We use all kinds of mechanisms to avoid confronting randomness. We talk about karma, in a cosmic equalization that ties seemingly unconnected events together. We believe in runs of luck, both good and ill, and that bad things happen in threes. We argue that

we are influenced by the stars, by the phases of the moon, by the motion of the planets in the heavens. When we get cancer, we automatically assume that something—or someone—is to blame.

But many events are not fully predictable or explicable. Disasters happen randomly, to good people as well as to bad ones, to star-crossed individuals as well as those who have a favorable planetary alignment. Sometimes you can make a good guess about the future, but randomness can confound even the most solid predictions: Don't be surprised when you're outlived by the overweight, cigar-smoking, speed-fiend motorcyclist down the block.

What's more, random events can mimic nonrandom ones. Even the most sophisticated scientists can have difficulty telling the difference between a real effect and a random fluke. Randomness can make placebos seem like miracle cures, or harmless compounds appear to be deadly poisons, and can even create subatomic particles out of nothing.

The Second Law of Randomness: Some events are impossible to predict. If you walk into a Las Vegas casino and observe the crowd gathered around the craps table, you'll probably see someone who thinks he's on a lucky streak. Because he's won several rolls in a row, his brain tells him he's going to keep winning, so he keeps gambling. You'll probably also see someone who's been losing. The loser's brain, like the winner's, tells him to keep gambling. Since he's been losing for so long, he thinks he's due for a stroke of luck; he won't walk away from the table, for fear of missing out.

Contrary to what our brains are telling us, there's no mystical force that imbues a winner with a streak of luck, nor is there a cosmic sense of justice that ensures that a loser's luck will turn around. The universe doesn't care one whit whether you've been winning or losing; each roll of the dice is just like every other.

No matter how much effort you put into observing how the dice have been behaving or how meticulously you have been watching for people who seem to have luck on their side, you get absolutely no information about what the next roll of a fair die will be. The outcome of a die roll is entirely independent of its history. And as a result, any scheme to gain some sort of advantage by observing the table is doomed to fail. Events like these—independent, purely random events—defy any attempts to find a pattern, because there is none to be found.

Randomness provides an absolute block against human ingenuity; it means that our logic, our science, our capacity for reason can penetrate only so far in predicting the behavior of the cosmos. Whatever methods you try, whatever theory you create, whatever logic you use to predict the next roll of a fair die, there's always a 5/6 chance you are wrong. Always.

The Third Law of Randomness: Random events behave predictably in aggregate even if they're not predictable individually.

Randomness is daunting; it sets limits where even the most sophisticated theories cannot go, shielding elements of nature from even our most determined inquiries. Nevertheless, to say that something is random is not equivalent to saying that we can't understand it. Far from it.

Randomness follows its own set of rules—rules that make the behavior of a random process understandable and predictable.

These rules state that even though a single random event might be completely unpredictable, a collection of independent random events is extremely predictable—and the larger the number of events, the more predictable they become. The law of large numbers is a mathematical theorem that dictates that repeated, independent random events converge with pinpoint accuracy upon a predictable average behavior. Another powerful mathematical tool, the central-limit theorem, tells you exactly how far off that average a given collection of events is likely to be. With these tools, no matter how chaotic, how strange, a random behavior might be in the short run, we can turn that behavior into stable, accurate predictions in the long run.

The rules of randomness are so powerful that they have given physics some of its most sacrosanct and immutable laws. Though the atoms in a box full of gas are moving at random, their collective behavior is described by a simple set of deterministic equations. Even the laws of thermodynamics derive their power from the predictability of large numbers of random events; they are indisputable only because the rules of randomness are so absolute.

Paradoxically, the unpredictable behavior of random events has given us the predictions that we are most confident in.

THE KALEIDOSCOPIC DISCOVERY ENGINE

CLIFFORD PICKOVER

Writer; associate editor, Computers and Graphics; *editorial board,* Odyssey, Leonardo, *and* YLEM; *author,* The Math Book: From Pythagoras to the 57th Dimension

The famous Canadian physician William Osler once wrote, "In science the credit goes to the man who convinced the world, not to the man to whom the idea first occurs." When we examine discoveries in science and mathematics, in hindsight we often find that if one scientist did not make a particular discovery, some other individual would have done so within a few months or years of the discovery. Most scientists, as Newton said, stood on the shoulders of giants to see the world just a bit farther along the horizon. Often, several people create essentially the same device or discover the same scientific law at about the same time, but for various reasons, including sheer luck, history sometimes remembers only one of them.

In 1858, the German mathematician August Möbius independently discovered the Möbius strip simultaneously with another German mathematician, Johann Benedict Listing. Isaac Newton and Gottfried Wilhelm Leibniz independently developed calculus at roughly the same time. British naturalists Charles Darwin and Alfred Russel Wallace both developed the theory of evolution by natural selection independently and simultaneously. Similarly, Hungarian mathematician János Bolyai and Russian mathematician Nikolai Lobachevsky seem to have developed hyperbolic geometry independently and at the same time.

The history of materials science is replete with simultaneous discoveries. For example, in 1886, the electrolytic process for refining aluminum using the mineral cryolite was discovered simultaneously and independently by American Charles Martin Hall and Frenchman Paul Héroult. Their inexpensive method for isolating pure aluminum from compounds had an enormous effect on industry. The time was "ripe" for such discoveries, given humanity's accumulated knowledge at the time the discoveries were made. On the other hand, mystics have suggested that a deeper meaning adheres to such coincidences. Austrian biologist Paul Kammerer wrote, "We thus arrive at the image of a world-mosaic or cosmic kaleidoscope, which, in spite of constant shufflings and rearrangements, also takes care of bringing like and like together." He compared events in our world to the tops of ocean waves that seem isolated and unrelated. According to his controversial theory, we notice the tops of the waves, but beneath the surface there may be some kind of synchronistic mechanism that mysteriously connects events in our world and causes them to cluster.

We are reluctant to believe that great discoveries are part of a discovery kaleidoscope and are mirrored in numerous individuals at once. However, as further examples, there were several independent discoveries of sunspots in 1611, even though Galileo gets most of the credit today. Alexander Graham Bell and Elisha Gray filed their own patents on telephone technologies on the same day. As the sociologist of science Robert Merton has remarked, "The genius is not a unique source of insight; he is merely an efficient source of insight."

Merton further suggested that "all scientific discoveries are in principle 'multiples.'" In other words, when a scientific discovery is made, it is made by more than one person. Sometimes a discovery is named after the person who develops the discovery, rather than after the original discoverer.

The world is full of difficulties in assigning credit for discoveries. Some of us have personally seen this in the realm of patent law, in business ideas, and in our daily lives. Fully appreciating the concept of the kaleidoscopic discovery engine adds to our cognitive toolkits because the kaleidoscope succinctly captures the nature of innovation and the future of ideas. If schools taught more about kaleidoscopic discovery, even in the context of everyday experience, then innovators might enjoy the fruits of their labor and still become "great" without a debilitating concern to be first or to crush rivals. The great eighteenth-century anatomist William Hunter frequently quarreled with his brother about who was first in making a discovery. But even Hunter admitted, "If a man has not such a degree of enthusiasm and love of the art, as will make him impatient of unreasonable opposition, and of encroachment upon his discoveries and his reputation, he will hardly become considerable in anatomy, or in any other branch of natural knowledge."

When Mark Twain was asked to explain why so many inventions were invented independently, he said, "When it's steamboat time, you steam."

INFERENCE TO THE BEST EXPLANATION

REBECCA NEWBERGER GOLDSTEIN
Philosopher; novelist; author, 36 Arguments for the Existence of God: A Work of Fiction

I'm alone in my home, working in my study, when I hear the click of the front door, the sound of footsteps making their way toward me. Do I panic? That depends on what I—my attention instantaneously appropriated to the task and cogitating at high speed—infer as the best explanation for those sounds. My husband returning home, the housecleaners, a miscreant breaking and entering, the noises of our old building settling, a supernatural manifestation? Additional details could make any one of these explanations, except the last, the best explanation for the circumstances.

Why not the last? As Charles Sanders Peirce, who first drew attention to this type of reasoning, pointed out: "Facts cannot be explained by a hypothesis more extraordinary than these facts themselves; and of various hypotheses the least extraordinary must be adopted."

"Inference to the best explanation" is ubiquitously pursued, which doesn't mean that it is ubiquitously pursued well. The phrase, coined by the Princeton philosopher Gilbert Harman as a substitute for Peirce's term, "abduction," should be in everybody's toolkit, if only because it forces one to think about what makes for a good explanation. There is that judgmental phrase, *the best*, sitting out in the open, shamelessly invoking standards. Not all explanations are created equal; some are objectively better than others. And the phrase also highlights another important fact. *The best* means the one that wins out over the alternatives, of which there are always many. Evidence calling for an explanation summons a great plurality (in fact, an infinity) of possible explanations, the great mass of which can be eliminated on the grounds of violating Peirce's maxim. We decide among the remainder using such criteria as: Which is the simpler, which does less violence to established beliefs, which is less ad hoc, which explains the most, which is the loveliest?

There are times when these criteria clash with one another. Inference to the best explanation is certainly not as rule-bound as logical deduction nor even as enumerative induction, which takes us from observed cases of all *a*'s being *b*'s to the probability that unobserved cases of *a*'s are also *b*'s. But inference to the best explanation also gets us a great deal more than either deduction or enumerative induction does.

It's inference to the best explanation that gives science the power to expand our ontology, giving us reasons to believe in things we can't directly observe, from subatomic particles—or maybe strings—to the dark matter and dark energy of cosmology. It's inference to the best explanation that allows us to know something of what it's like to be other people on the basis of their behavior. I see the hand drawing too near the fire and then quickly pull away, tears starting in the eyes while an impolite word is uttered, and I know something of what that person is feeling. It's on the basis of inference to the best explanation that I can learn things from what authorities say and write, my inferring that the best explanation for their doing so is that they believe what they say or write. (Sometimes that's not the best explanation.) In fact, I'd argue that my right to believe in a world outside my own solipsistic universe, confined to the awful narrowness of my own immediate experience, is based on inference to the best explanation. What best explains the vivacity and predictability of some of my representations of material bodies and not others, if not the hypothesis of actual material bodies? Inference to the best explanation defeats mental-sapping skepticism.

Many of our most rancorous scientific debates—say, over string theory or the foundations of quantum mechanics—have been over which competing criteria for judging explanations *the best* ought to prevail. So, too, have debates that many of us have conducted over scientific versus religious explanations. These debates could be sharpened by bringing to bear on them the rationality-steeped notion of inference to the best explanation, its invocation of the sorts of standards that make some explanations objectively better than others, beginning with Peirce's enjoinder that extraordinary hypotheses be ranked far away from the best.

PRAGMAMORPHISM

EMANUEL DERMAN
Professor of financial engineering, Columbia University; principal, Prisma Capital Partners; former head, Quantitative Strategies Group, Equities Division, Goldman Sachs & Co.; author, My Life as a Quant: Reflections on Physics and Finance

Anthropomorphism means attributing the characteristics of human beings to inanimate things or animals. I have invented the word "pragmamorphism" as a shorthand abstraction for the attribution of the properties of inanimate things to human beings. One of the meanings of the Greek word *pragma* is "a material object."

Being pragmamorphic sounds equivalent to taking a scientific attitude toward the world, but it easily evolves into dull scientism. It's pragmamorphic to equate material correlates with human psychological states—to equate PET scans, say, with emotion. It's also pragmamorphic to ignore human qualities you cannot measure.

We have discovered useful metrics for material objects—length, temperature, pressure, volume, kinetic energy, etc. "Pragmamorphism" is a good word for the attempt to assign such one-dimensional thing-metrics to the mental qualities of humans. IQ, a "length scale" for intelligence, is a result of pragmamorphism. But intelligence is more diffuse than linear.

The utility function in economics is similar. It's clear that people have preferences. But is it clear that there is a function that describes their preferences?

COGNITIVE LOAD

NICHOLAS CARR
Science and technology journalist; author, The Shallows: What the Internet Is Doing to Our Brains

You're sprawled on the couch in your living room, watching a new episode of *Justified* on the tube, when you think of something you need to do in the kitchen. You get up, take ten quick steps across the carpet, and then just as you reach the kitchen—*poof!*—you realize you've forgotten what it was you got up to do. You stand befuddled for a moment, then shrug your shoulders and head back to the couch.

Such memory lapses happen so often that we don't pay them much heed. We write them off as "absentmindedness" or, if we're getting older, "senior moments." But the incidents reveal a fundamental limitation of our minds: the tiny capacity of our working memory. Working memory is what brain scientists call the short-term store of information where we hold the contents of our consciousness at any given moment—all the impressions and thoughts that flow into our mind as we go through a day. In the 1950s, Princeton psychologist George Miller famously argued that our brains can hold only about seven pieces of information simultaneously. Even that figure may be too high. Some brain researchers now believe that working memory has a maximum capacity of just three or four elements.

The amount of information entering our consciousness at any instant is referred to as our cognitive load. When our cognitive load exceeds the capacity of our working memory, our intellectual abilities take a hit. Information zips into and out of our mind so quickly that we never gain a good mental grip on it. (Which is why you can't remember what you went to the kitchen to do.) The information vanishes before we've had an opportunity to transfer it into our long-term memory and weave it into knowledge. We remember less, and our ability to think critically and conceptually weakens. An overloaded working memory also tends to increase our distractedness. After all, as the neuroscientist Torkel Klingberg has pointed out, "We have to remember what it is we are to concentrate on." Lose your hold on that and you'll find "distractions more distracting."

Developmental psychologists and educational researchers have long used the concept of cognitive load in designing and evaluating pedagogical techniques. They know that when you give a student too much information too quickly, comprehension degrades and learning suffers. But now that all of us—thanks to the incredible speed and volume of modern digital communication networks and gadgets—are inundated with more bits and pieces of information than ever before, everyone would benefit from having an understanding of cognitive load and how it influences memory and thinking. The more aware we are of how small and fragile our working memory is, the better we'll be able to monitor and manage our cognitive load. We'll become more adept at controlling the flow of the information coming at us.

There are times when you want to be awash in messages and other info-bits. The resulting sense of connectedness and stimulation can be exciting and pleasurable. But it's important to remember that when it comes to the way your brain works, information overload is not just a metaphor, it's a physical state. When you're engaged in a particularly important or complicated intellectual task, or when you simply want to savor an experience or a conversation, it's best to turn the information faucet down to a trickle.

TO CURATE

HANS ULRICH OBRIST
Curator, Serpentine Gallery, London

Lately, the term "to curate" seems to be used in a greater variety of contexts than ever before, in reference to everything from an exhibition of prints by old masters to the contents of a concept store. The risk, of course, is that the definition may expand beyond functional usability. But I believe "to curate" finds ever wider application because of a feature of modern life impossible to ignore: the incredible proliferation of ideas, information, images, disciplinary knowledge, and material products we all witness today. Such proliferation makes the activities of filtering, enabling, synthesizing, framing, and remembering more and more important as basic navigational tools for twenty-first-century life. These are the tasks of the curator, who is no longer understood simply as the person who fills a space with objects but also as the person who brings different cultural spheres into contact, invents new display features, and makes junctions that allow unexpected encounters and results.

Michel Foucault once wrote that he hoped his writings would be used by others as a theoretical toolbox, a source of concepts and models for understanding the world. For me, the author, poet, and theoretician Édouard Glissant has become this kind of toolbox. Very early, he noted that in our phase of globalization (which is not the first one), there is a danger of homogenization but at the same time a countermovement, the retreat into one's own culture. And against both dangers he proposes the idea of *mondialité*—a global dialog that *augments* difference.

This inspired me to handle exhibitions in a new way. There

is a lot of pressure on curators not only to do shows in one place but also to send them around the world, by simply packing them into boxes in one city and unpacking them in the next. This is a homogenizing sort of globalization. Using Glissant's idea as a tool means to develop exhibitions that build a relation to their place, that change with their different local conditions, that create a changing dynamic complex system with feedback loops.

To curate in this sense is to refuse static arrangements and permanent alignments and instead to enable conversations and relations. Generating these kinds of links is an essential part of what it means to curate, as is disseminating new knowledge, new thinking, and new artworks in a way that can seed future cross-disciplinary inspirations.

But there is another case for curating as a vanguard activity for the twenty-first century. As the artist Tino Sehgal has pointed out, modern societies find themselves today in an unprecedented situation: The problem of lack, or scarcity, which has been the primary factor motivating scientific and technological innovation, is now joined and even superseded by the problem of the global effects of overproduction and resource use. Thus, moving beyond the object as the locus of meaning has a further relevance. Selection, presentation, and conversation are ways for human beings to create and exchange real value, without dependence on older, unsustainable processes. Curating can take the lead in pointing us toward this crucial importance of choosing.

"GRACEFUL" SHAS

RICHARD NISBETT
Social psychologist; codirector, Culture and Cognition Program, University of Michigan; author, Intelligence and How to Get It: Why Schools and Cultures Count

1. A university needs to replace its aging hospital. Cost estimates indicate that it would be as expensive to remodel the old hospital as to demolish it and build a new one from scratch. The main argument offered by the proponents of the former is that the original hospital was very expensive to build and it would be wasteful to simply demolish it. The main argument of the proponents of a new hospital is that a new hospital would inevitably be more modern than a remodeled one. Which seems wiser to you—remodel or build a new hospital?

2. David L., a high school senior, is choosing between two colleges, equal in prestige, cost, and distance from home. David has friends at both colleges. Those at College A like it from both intellectual and personal standpoints. Those at College B are generally disappointed on both grounds. But David visits each college for a day, and his impressions are quite different from those of his friends. He meets several students at College A who seem neither particularly interesting nor particularly pleasant, and a couple of professors give him the brush-off. He meets several bright and pleasant students at College B and two professors take a personal interest in him. Which college do you think David should go to?

3. Which of the cards below should you turn over to answer to determine whether the following rule has been violated or not? "If there is a vowel on the front of the card, then there is an odd number on the back."

U	**K**	**3**	**8**

Some considerations about each of these questions:

Question 1: If you said that the university should remodel on the grounds that it was expensive to build the old hospital, you have fallen into the "sunk-cost trap" shorthand abstraction (SHA) identified by economists. The money spent on the hospital is irrelevant—it's sunk—and has no bearing on the present choice. Amos Tversky and Daniel Kahneman pointed out that people's ability to avoid such traps might be helped by a couple of thought experiments like the following:

Imagine that you have two tickets to tonight's NBA game in your city and that the arena is forty miles away. But it's begun to snow, and you've found out that your team's star has been injured and won't be playing. Should you go, or just throw away the money and skip it?

To answer that question as an economist would, ask yourself the following question: "Suppose you didn't have tickets to the game, and a friend called you up and said he had two tickets to tonight's game which he couldn't use, and asked if you'd like to have them." If the answer is "You've got to be kidding. It's snowing and the star isn't playing," then the answer is that you shouldn't go. That answer shows you that the fact that you paid good money for the tickets you have is irrelevant—their cost is sunk and can't be retrieved by doing something you don't want to do anyway. Avoidance of sunk-cost traps is a religion for economists, but I find that a single college course in economics actually does little to make people aware of the sunk-cost trap. It turns out that exposure to a few basketball-type anecdotes does a lot.

Question 2: If you said that "David is not his friends; he should go to the place he likes," then the SHA of the law of large numbers (LLN) has not been sufficiently salient to you. David has one day's worth of experiences at each; his friends have had hundreds. Unless David thinks his friends have kinky tastes, he should ignore his own impressions and go to College A. A single college course in statistics increases the likelihood of invoking LLN. Several courses in statistics make LLN considerations almost inevitable.

Question 3: If you said anything other than "Turn over the U and turn over the 8," psychologists P. C. Wason and P. N. Johnson-Laird have shown that you would be in the company of 90 percent of Oxford students. Unfortunately, you—and they—are wrong. The SHA of the logic of the conditional has not guided your answer. "If P, then Q" is satisfied by showing that the P is associated with a Q and the not-Q is not associated with a P. A course in logic actually does nothing to make people better able to answer questions such as number 3. Indeed, a PhD in philosophy does nothing to make people better able to apply the logic of the conditional to simple problems like Question 3 or meatier problems of the kind one encounters in everyday life.

Some SHAs apparently are "graceful," in that they are easily inserted into the cognitive toolbox. Others appear to be clunky and don't readily fit. If educators want to improve people's ability to think, they need to know which SHAs are graceful and teachable and which are clunky and hard to teach. An assumption of educators for centuries has been that formal logic improves thinking skills—meaning that it makes people more intelligent in their everyday lives. But this belief may be mistaken. (Bertrand Russell said, almost surely correctly, that the syllogisms studied by the monks of medieval Europe were as sterile as they were.) But it seems likely that many crucially important SHAs, undoubtedly including some that have been proposed by this year's *Edge* contributors, are readily taught. Few questions are more important for educators to study than to find out which SHAs are teachable and how they can be taught most easily.

EXTERNALITIES

ROB KURZBAN
Psychologist, University of Pennsylvania; director, Pennsylvania Laboratory for Experimental Evolutionary Psychology (PLEEP); author, Why Everyone (Else) Is a Hypocrite: Evolution and the Modular Mind

When I go about doing what I do, frequently I affect you as an incidental side effect. In many such cases, I don't have to pay you to compensate for any inadvertent harm done; symmetrically, you frequently don't have to pay me for any inadvertent benefits I've bestowed upon you. The term "externalities" refers to these cases, and they are pervasive and important because, especially in the modern, interconnected world, when I pursue my own goals I wind up affecting you in any number of different ways.

Externalities can be small or large, negative and positive. When I lived in Santa Barbara, many people with no goal other than working on their tans generated (small, it's true) positive externalities for passersby, who benefited from the enhanced scenery. These onlookers didn't have to pay for this improvement to the landscape, but at the same beach Rollerbladers traveling at high speed and distracted by this particular positive externality occasionally produced a negative one, in the form of a risk of collision for pedestrians trying to enjoy the footpath.

Externalities are increasingly important in the present era, when actions in one place potentially affect others half a world away. When I manufacture widgets for you to buy, I might, as a side effect of the process, produce waste that makes the people around my factory—and maybe around the world—worse off. As long as I don't have to compensate anyone for polluting their water and air, it's unlikely I'll make much of an effort to stop doing it.

On a smaller, more personal scale, we all impose externalities on one another as we go through our daily lives. I drive to work, increasing the amount of traffic you face. You feel the strange compulsion that infects people in theaters these days to check your text messages on your cell phone during the film, and the bright glow peeking over your shoulder reduces my enjoyment of the movie.

The concept of externalities is useful because it directs our attention to such unintended side effects. If you weren't focused on externalities, you might think that the way to reduce traffic congestion was to build more roads. That might work, but another way, and a potentially more efficient way, is to implement policies that force drivers to pay the cost of their negative externalities by charging a fee to use roads, particularly at peak times. Congestion charges, such as those implemented in London and Singapore, are designed to do exactly that. If I have to pay to go into town during rush hour, I may stay home unless my need is pressing.

Keeping externalities firmly in mind also reminds us that in complex, integrated systems, simple interventions designed to bring about a particular desirable effect will potentially have many more consequences, both positive and negative. Consider, as an example, the history of DDT. When first used, it had its intended effect, which was to reduce the spread of malaria through the control of mosquito populations. However, its use also had two unintended consequences. First, it poisoned a number of animals (including humans), and, second, it selected for resistance among mosquitoes. Subsequently, policies to reduce the use of DDT probably were effective in preventing these two negative consequences. However, while there is some debate about the details, these policies might themselves have had an important side effect—increasing rates of malaria, carried by the mosquitoes no longer suppressed by DDT.

The key point is that the notion of externalities forces us to think about unintended (positive and negative) effects of actions, an issue that looms larger as the world gets smaller. It highlights the need to balance not only the intended costs and benefits of a given candidate policy but also its unintended effects. Further, it helps us focus on one type of solution to the problems of unintended harms, which is to think about using financial incentives for people and firms to produce more positive externalities and fewer negative ones.

Considering externalities in our daily lives directs our attention to ways in which we harm, albeit inadvertently, the people around us, and can guide our decision making—including waiting until after the credits have rolled to check our messages.

EVERYTHING IS IN MOTION

JAMES O'DONNELL
Classicist; provost, Georgetown University; author,
The Ruin of the Roman Empire

Nothing is more wonderful about human beings than their ability to abstract, infer, calculate, and produce rules, algorithms, and tables that enable them to work marvels. We are the only species that could even imagine taking on Mother Nature in a fight for control of the world. We may well lose that fight, but it's an amazing spectacle nonetheless.

But nothing is less wonderful about human beings than their ability to refuse to learn from their own discoveries. The edge to the *Edge* Question this year is the implication that we are brilliant and stupid at the same time, capable of inventing wonders and still capable of forgetting what we've done and blundering stupidly on. Our poor cognitive toolkits are always missing a screwdriver when we need one, and we're always trying to get a bolt off that wheel with our teeth, when a perfectly serviceable wrench is in the kit over there, unused.

So as a classicist, I'll make my pitch for what is arguably the oldest of our SHA concepts, the one that goes back to the senior pre-Socratic philosopher Heraclitus. "You can't step in the same river twice," he said. Putting it another way, his mantra was "Everything flows." Remembering that everything is in motion—feverish, ceaseless, unbelievably rapid motion—is always hard for us. Vast galaxies dash apart at speeds that seem faster than is physically possible, while the subatomic particles of which we are composed beggar our ability to comprehend large numbers when we try to understand their motion—and at the same time, I lie here, sluglike, inert, trying to muster the energy to change channels, convinced that one day is just like another.

Because we think and move on a human scale in time and space, we can deceive ourselves. Pre-Copernican astronomies depended on the self-evident fact that the "fixed stars" orbited the Earth in a slow annual dance; and it was an advance in science to declare that "atoms" (in Greek, "indivisibles") were the changeless building blocks of matter—until we split them. Edward Gibbon was puzzled by the fall of the Roman Empire because he failed to realize that its most amazing feature was that it lasted so long. Scientists discover magic disease-fighting compounds only to find that the disease changes faster than they can keep up.

Take it from Heraclitus and put it in your toolkit: Change is the law. Stability and consistency are illusions, temporary in any case, a heroic achievement of human will and persistence at best. When we want things to stay the same, we'll always wind up playing catch-up. Better to go with the flow.

SUBSELVES AND THE MODULAR MIND

DOUGLAS T. KENRICK
Professor of social psychology, Arizona State University; author,
Sex, Murder, and the Meaning of Life

Although it seems obvious that there is a single "you" inside your head, research from several subdisciplines of psychology suggests that this is an illusion. The "you" who makes a seemingly rational and "self-interested" decision to discontinue a relationship with a friend who fails to return your phone calls, borrows thousands of dollars he doesn't pay back, and lets you pick up the tab in the restaurant is not the same "you" who makes very different calculations about a son, a lover, or a business partner.

Three decades ago, cognitive scientist Colin Martindale advanced the idea that each of us has several subselves, and he connected his idea to emerging ideas in cognitive science. Central to Martindale's thesis were a few fairly simple ideas, such as selective attention, lateral inhibition, state-dependent memory, and cognitive dissociation. Although there are billions of neurons in our brains firing all the time, we'd never be able to put one foot in front of the other if we were unable to ignore almost all of that hyperabundant parallel processing going on in the background. When you walk down the street, there are thousands of stimuli to stimulate your already overtaxed brain—hundreds of different people of different ages with different accents, different hair colors, different clothes, different ways of walking and gesturing, not to mention all the flashing advertisements, curbs to avoid tripping over, and automobiles running yellow lights as you try to cross at the intersection. Hence, attention is highly selective. The nervous system accomplishes some of that selectiveness by relying on the powerful principle of lateral inhibition—in which one group of neurons suppresses the activity of other neurons that might interfere with an important message getting up to the next level of processing. In the eye, lateral inhibition helps us notice potentially dangerous holes in the ground, as the retinal cells stimulated by light areas send messages suppressing the activity of neighboring neurons, producing a perceived bump in brightness and valley of darkness near any edge. Several of these local "edge detector"–style mechanisms combine at a higher level to produce "shape detectors," allowing us to discriminate a "b" from a "d" and a "p." Higher up in the nervous system, several shape detectors combine to allow us to discriminate words, and at a higher level to discriminate sentences, and at a still higher level to place those sentences in context (thereby determining whether the statement "Hi, how are you today?" is a romantic pass or a prelude to a sales pitch).

State-dependent memory helps sort out all that incoming information for later use by categorizing new info according to context: If you learn a stranger's name after drinking a doppio espresso with her at the local java house, it will be easier to remember that name if you meet again at Starbucks than if the next encounter is at a local pub after a martini. For several months after I returned from Italy, I would start speaking Italian and making expansive hand gestures every time I drank a glass of wine.

Martindale argued that at the highest level all those processes of inhibition and dissociation lead us to suffer from an everyday version of dissociative disorder. In other words, we all have a number of executive subselves, and the only way we manage to accomplish anything in life is to allow only one subself to take the conscious driver's seat at any given time.

Martindale developed his notion of executive subselves before modern evolutionary approaches to psychology had become prominent, but the idea becomes especially powerful if you combine his cognitive model with the idea of *functional modularity*. Building on findings that animals and humans use remarkably varied mental processes to learn different things, evolutionarily informed psychologists have suggested that there is not a single information-processing organ inside our heads but instead multiple systems dedicated to solving different adaptive problems. Thus, instead of having a random and idiosyncratic assortment of subselves inside my head unlike the assortment inside your head, each of us has a set of functional subselves—one dedicated to getting along with our friends, one dedicated to self-protection (protecting us from the bad guys), one dedicated to winning status, another to finding mates, yet another to keeping mates (which presents a very different set of problems, as some of us have learned), and still another to caring for our offspring.

Thinking of the mind as composed of several functionally independent adaptive subselves helps us understand many apparent inconsistencies and irrationalities in human behavior, such as why a decision that seems rational when it involves one's son seems eminently irrational when it involves a friend or a lover, for example.

PREDICTIVE CODING

ANDY CLARK

Professor of philosophy, University of Edinburgh; author, Supersizing the Mind: Embodiment, Action, and Cognitive Extension

The idea that the brain is basically an engine of prediction is one that will, I believe, turn out to be very valuable not just within its current home (computational cognitive neuroscience) but across the board—for the arts, for the humanities, and for our own personal understanding of what it is to be a human being in contact with the world.

The term "predictive coding" is currently used in many ways, across a variety of disciplines. The usage I recommend for the Everyday Cognitive Toolkit is, however, more restricted in scope. It concerns the way the brain exploits prediction and anticipation in making sense of incoming signals and using them to guide perception, thought, and action. Used in this way, predictive coding names a technically rich body of computational and neuroscientific research (key theorists include Dana Ballard, Tobias Egner, Paul Fletcher, Karl Friston, David Mumford, and Rajesh Rao). This corpus of research uses mathematical principles and models that explore in detail the ways that this form of coding might underlie perception and inform belief, choice, and reasoning.

The basic idea is simple. It is that to perceive the world is to successfully predict our own sensory states. The brain uses stored knowledge about the structure of the world and the probabilities of one state or event following another to generate a prediction of what the current state is likely to be, given the previous one and this body of knowledge. Mismatches between the prediction and the received signal generate error signals that nuance the prediction or (in more extreme cases) drive learning and plasticity.

We may contrast this with older models in which perception is a "bottom-up" process, in which incoming information is progressively built (via some kind of evidence-accumulation process, starting with simple features and working up) into a high-level model of the world. According to the predictive-coding alternative, the reverse is the case. For the most part, we determine the low-level features by applying a cascade of predictions that begin at the very top—with our most general expectations about the nature and state of the world providing constraints on our successively more detailed (fine-grain) predictions.

This inversion has some quite profound implications.

First, the notion of good ("veridical") sensory contact with the world becomes a matter of applying the right expectations to the incoming signal. Subtract such expectations and the best we can hope for are prediction errors that elicit plasticity and learning. This means, in effect, that all perception is some form of "expert perception," and that the idea of accessing some kind of unvarnished sensory truth is untenable (unless that merely names another kind of trained, expert perception!).

Second, the time course of perception becomes critical. Predictive-coding models suggest that what emerges first is the general gist (including the general affective feel) of the scene, with the details becoming progressively filled in as the brain uses that larger context—time and task allowing—to generate finer and finer predictions of detail. There is a very real sense in which we properly perceive the forest before the trees.

Third, the line between perception and cognition becomes blurred. What we perceive (or think we perceive) is heavily determined by what we know, and what we know (or think we know) is constantly conditioned on what we perceive (or think we perceive). This turns out to offer a powerful window on various pathologies of thought and action, explaining the way hallucinations and false beliefs go hand in hand in schizophrenia, as well as other more familiar states such as "confirmation bias" (our tendency to "spot" confirming evidence more readily than disconfirming evidence).

Fourth, if we now consider that prediction errors can be suppressed not just by changing predictions but also by changing the things predicted, we have a simple and powerful explanation for behavior and the way we manipulate and sample our environment. In this view, action is there to make predictions come true and provides a nice account of phenomena that range from homeostasis to the maintenance of our emotional and interpersonal status quo.

Understanding perception as prediction thus offers, it seems to me, an excellent tool for appreciating both the power and the potential hazards of our primary way of being in contact with the world. Our primary contact with the world, all this suggests, is via our expectations about what we're about to see or experience. The notion of predictive coding, by offering a concise and technically rich way of gesturing at this fact, provides a cognitive tool that will more than earn its keep in science, law, ethics, and the understanding of our own daily experience.

OUR SENSORY DESKTOP

DONALD HOFFMAN
Cognitive scientist, University of California–Irvine; author, Visual Intelligence: How We Create What We See

Our perceptions are neither true nor false. Instead, our perceptions of space and time and objects—the fragrance of a rose, the tartness of a lemon—are all part of our "sensory desktop," which functions much like a computer desktop.

Graphical desktops for personal computers have existed for about three decades. Yet they are now such an integral part of daily life that we might easily overlook a useful concept that they embody. A graphical desktop is a *guide to adaptive behavior*. Computers are notoriously complex devices, more complex than most of us care to learn. The colors, shapes, and locations of icons on a desktop shield us from the computer's complexity, and yet they allow us to harness its power by appropriately informing our behaviors, such as mouse movements and button clicks that open, delete, and otherwise manipulate files. In this way, a graphical desktop is a guide to adaptive behavior.

Graphical desktops make it easier to grasp the idea that guiding adaptive behavior is different from reporting truth. A red icon on a desktop does not report the true color of the file it represents. Indeed, a file has no color. Instead, the red color guides adaptive behavior, perhaps by signaling the relative importance or recent updating of the file. The graphical desktop guides useful behavior and hides what is true but not useful. The complex truth about the computer's logic gates and magnetic fields is, for the purposes of most users, of no use.

Graphical desktops thus make it easier to grasp the nontrivial difference between utility and truth. Utility drives evolution by natural selection. Grasping the distinction between utility and truth is therefore critical to understanding a major force that shapes our bodies, minds, and sensory experiences.

Consider, for instance, facial attractiveness. When we glance at a face, we get an immediate sense of its attractiveness, an impression that usually falls somewhere between hot and not. That feeling can inspire poetry, evoke disgust, or launch a thousand ships. It certainly influences dating and mating. Research in evolutionary psychology suggests that this feeling of attractiveness is a guide to adaptive behavior. The behavior is mating, and the initial feeling of attraction toward a person is an adaptive guide because it correlates with the likelihood that mating with that person will lead to successful offspring.

Just as red does not report the true color of a file, so hotness does not report the true attraction of a face: Files have no intrinsic color; faces have no intrinsic attraction. The color of an icon is an artificial convention to represent aspects of the utility of a colorless file. The initial sense of attractiveness is an artificial convention to represent mate utility.

The phenomenon of synesthesia can help us to understand the conventional nature of our sensory experiences. In many cases of synesthesia, a stimulus that is normally experienced in one way (say, as a sound) is also automatically experienced in another way (say, as a color). Someone with sound-color synesthesia sees colors and simple shapes whenever they hear a sound. The same sound always occurs with the same colors and shapes. People with taste-touch synesthesia feel touch sensations in their hands every time they taste something. The same taste always occurs with the same feeling of touch in their hands. The particular connections between sound and color that one sound-color synesthete experiences typically differ from the connections experienced by another such synesthete. In this sense, the connections are an arbitrary convention. Now, imagine a sound-color synesthete who no longer has sound experiences from acoustic stimuli and instead has only the synesthetic color experiences. This synesthete would experience only as colors what the rest of us experience as sounds. In principle, they could get all the acoustic information the rest of us get, except in a color format rather than a sound format.

This leads to the concept of a sensory desktop. Our sensory experiences—such as vision, sound, taste, and touch—can be thought of as sensory desktops that have evolved to guide adaptive behavior, not report objective truths. As a result, we should take our sensory experiences seriously. If something tastes putrid, we probably shouldn't eat it. If it sounds like a rattlesnake, we probably should avoid it. Our sensory experiences have been shaped by natural selection to guide such adaptive behaviors.

We must take our sensory experiences seriously but not literally. This is one place where the concept of a sensory desktop is helpful. We take the icons on a graphical desktop seriously; we won't, for instance, carelessly drag an icon to the trash, for fear of losing a valuable file. But we don't take the colors, shapes, or locations of the icons literally. They are not there to resemble the truth. They are there to facilitate useful behaviors.

Sensory desktops differ across species. A face that could launch a thousand ships probably has no attraction to a macaque. The carrion that tastes putrid to me might taste like a delicacy to a vulture. My taste experience guides behaviors appropriate for me; eating carrion could kill me. The vulture's taste experience guides behaviors appropriate to it; carrion is its primary food source.

Much of evolution by natural selection can be understood as an arms race between competing sensory desktops. Mimicry and camouflage exploit limitations in the sensory desktops of predators and prey. A mutation that alters a sensory desktop to reduce such exploitation conveys a selective advantage. This cycle of exploiting and revising sensory desktops is a creative engine of evolution.

On a personal level, the concept of a sensory desktop can enhance our cognitive toolkit by refining our attitude toward our own perceptions. It is common to assume that the way I see the world is, at least in part, the way it really is. Because, for instance, I experience a world of space and time and objects, it is common to assume that these experiences are, or at least resemble, objective truths. The concept of a sensory desktop reframes all this. It loosens the grip of sensory experiences on the imagination. Space, time, and objects might just be aspects of a sensory desktop specific to *Homo sapiens*. They might not be deep insights into objective truths, just convenient conventions that have evolved to allow us to survive in our niche. Our desktop is just a desktop.

THE SENSES AND THE MULTISENSORY

BARRY C. SMITH
Director, Institute of Philosophy, School of Advanced Study, University of London; writer and presenter, BBC World Service series The Mysteries of the Brain

For far too long, we have labored under a faulty conception of the senses. Ask anyone you know how many senses we have and they will probably say five—unless they start talking to you about a sixth sense. But why pick five? What of the sense of balance provided by the vestibular system, telling you whether you are going up or down in a lift, forward or backward on a train, or side to side on a boat? What about proprioception that gives you a firm sense of where your limbs are when you close your eyes? What about feeling pain, or heat and cold? Are these just part of touch, like feeling velvet or silk? And why think of sensory experiences like seeing, hearing, tasting, touching, and smelling as being produced by a single sense?

Contemporary neuroscientists have postulated two visual systems—one responsible for how things look to us, the other for controlling action—that operate independently of each other. The eye may fall for visual illusions, but the hand does not, reaching smoothly for a shape that looks larger than it is.

And it doesn't stop here. There is good reason to think that we have two senses of smell: (1) an external sense of smell—orthonasal olfaction, produced by inhaling—that enables us to detect such things in the environment as food, predators, and smoke; and (2) an internal sense—retronasal olfaction, produced by exhaling—

that enables us to detect the quality of what we have just eaten, allowing us to decide whether we want any more or should expel it. Associated with each sense of smell is a distinct hedonic response. Orthonasal olfaction gives rise to the pleasure of anticipation. Retronasal olfaction gives rise to the pleasure of reward. Anticipation is not always matched by reward. Have you ever noticed how the enticing aromas of freshly brewed coffee are never quite matched by the taste? There is always a little disappointment. Interestingly, the one food where the intensity of orthonasally and retronasally judged aromas match perfectly is chocolate. We get just what we expected, which may explain why chocolate is such a powerful stimulus.

Besides the proliferation of the senses in contemporary neuroscience, another major change is taking place. We used to study the senses in isolation, with the greatest majority of researchers focusing on vision. Things are rapidly changing. We now know that the senses do not operate in isolation but combine, both at early and late stages of processing, to produce our rich perceptual experiences of our surroundings. It is almost never the case that our experience presents us with just sights or sounds. We are always enjoying conscious experiences made up of sights and sounds, smells, the feel of our body, the taste in our mouths—and yet these are not presented as separate sensory parcels. We simply take in the rich and complex scene without giving much thought to how the different contributors produce the whole experience.

We give little thought to how smell provides a background to every conscious waking moment. People who lose their sense of smell can be plunged into depression and, a year later, show less sign of recovery than people who lose their sight. This is because familiar places no longer smell the same and people no longer have their reassuring olfactory signature. Also, patients who lose their smell believe they have lost their sense of taste. When tested, they acknowledge that they can taste sweet, sour, salt, bitter, savory, and metallic. But everything else, missing from the taste of what they are eating, is due to retronasal smell.

What we call "taste" is one of the most fascinating case studies for how inaccurate our view of our senses is: It is not produced by the tongue alone but is always an amalgam of taste, touch, and smell. Touch contributes to sauces tasting creamy and other foods tasting chewy, crisp, or stale. The only difference between potato chips that "taste" fresh or stale is a difference in texture. The largest part of what we call "taste" is in fact smell in the form of retronasal olfaction, which is why people who lose their ability to smell say they can no longer taste anything. Taste, touch, and smell are not merely combined to produce experiences of foods or liquids; rather, the information from the separate sensory channels is fused into a unified experience of what we call taste and what food scientists call *flavor*.

Flavor perception is the result of multisensory integration of gustatory, olfactory, and oral somatosenory information into a single experience whose components we are unable to distinguish. It is one of the most multisensory experiences we have, and it can be influenced by both sight and sound. The colors of wines and the sounds that food make when we bite or chew them can significantly affect our resulting appreciation and assessment; irritation of the trigeminal nerve in the face will make chilies feel "hot" and menthol feel "cool" in the mouth without any actual change in temperature.

In sensory perception, multisensory integration is the rule, not the exception. In audition, we don't just hear with our ears, we use our eyes to locate the apparent sources of sounds in the cinema where we "hear" the voices coming from the actors' mouths on the screen, although the sounds are coming from the sides of the theater. This is known as the ventriloquism effect. Similarly, retronasal odors detected by olfactory receptors in the nose are experienced as tastes in the mouth. The sensations get relocated to the mouth because oral sensations of chewing or swallowing capture our attention, making us think these olfactory experiences are occurring in the same place.

Other surprising collaborations among the senses are due to cross-modal effects, whereby stimulation of one sense boosts activity in another. Looking at someone's lips across a crowded room can improve our ability to hear what they are saying, and the smell of vanilla can make a liquid we sip "taste" sweeter and less sour. This is why we say vanilla is sweet-smelling, although sweet is a taste and pure vanilla is not sweet at all. Industrial manufacturers know about these effects and exploit them. Certain aromas in shampoos, for example, can make the hair "feel" softer; red-colored drinks "taste" sweet, whereas drinks with a light green color "taste" sour. In many of these interactions vision will dominate, but not in every case.

People unlucky enough to have a disturbance in their vestibular system will feel the world is spinning, although cues from the eyes and body should be telling them that everything is still. In people without this difficulty, the brain goes with the vision, and proprioception falls into line. Luckily, our senses cooperate, and we get ourselves around the world we inhabit—and that world is not a sensory but a multisensory world.

THE UMWELT

DAVID EAGLEMAN

Neuroscientist; director, Laboratory for Perception and Action, Initiative on Neuroscience and Law, Baylor College of Medicine; author, Incognito: The Secret Lives of the Brain

In 1909, the biologist Jakob von Uexküll introduced the concept of the *umwelt*. He wanted a word to express a simple (but often overlooked) observation: Different animals in the same ecosystem pick up on different environmental signals. In the blind and deaf world of the tick, the important signals are temperature and the odor of butyric acid. For the black ghost knifefish, it's electrical fields. For the echolocating bat, it's air-compression waves. The small subset of the world that an animal is able to detect is its *umwelt*. The bigger reality, whatever that might mean, is called the *umgebung*.

The interesting part is that each organism presumably assumes its umwelt to be the entire objective reality "out there." Why would any of us stop to think that there is more beyond what we can sense? In the movie *The Truman Show*, the eponymous Truman lives in a world completely constructed around him by an intrepid television producer. At one point, an interviewer asks the producer, "Why do you think Truman has never come close to discovering the true nature of his world?" The producer replies, "We accept the reality of the world with which we're presented." We accept our umwelt and stop there.

To appreciate the amount that goes undetected in our lives, imagine you're a bloodhound. Your long nose houses 200 million scent receptors. On the outside, your wet nostrils attract and trap scent molecules. The slits at the corners of each nostril flare out to allow more air flow as you sniff. Even your floppy ears drag along the ground and kick up scent molecules. Your world is all about olfaction. One afternoon, as you're following your master, you're stopped in your tracks by a revelation. What is it like to have the pitiful, impoverished nose of a human being? What can humans possibly detect when they take in a feeble little noseful of air? Do they suffer a hole where smell is supposed to be?

Obviously, we suffer no absence of smell, because we accept reality as it's presented to us. Without the olfactory capabilities of a bloodhound, it rarely strikes us that things could be different. Similarly, until a child learns in school that honeybees enjoy ultraviolet signals and rattlesnakes employ infrared, it does not strike her that plenty of information is riding on channels to which we have no natural access. From my informal surveys, it is very *un*common knowledge that the part of the electromagnetic spectrum visible to us is less than a ten-trillionth of it.

A good illustration of our unawareness of the limits of our umwelt is that of color-blind people: Until they learn that others can see hues they cannot, the thought of extra colors does not hit their radar screen. The same goes for the congenitally blind: Being sightless is not like experiencing "blackness" or "a dark hole" where vision should be. Like the human compared with the bloodhound, blind people do not miss vision; they do not conceive of it. The visible part of the spectrum is simply not part of their umwelt.

The more science taps into these hidden channels, the more it becomes clear that our brains are tuned to detect a shockingly small fraction of the surrounding reality. Our sensorium is enough for us to get by in our ecosystem, but it does not approximate the larger picture.

It would be useful if the concept of the umwelt were embedded in the public lexicon. It neatly captures the idea of limited knowledge, of unobtainable information, of unimagined possibilities. Consider the criticisms of policy, the assertions of dogma, the declarations of fact that you hear every day, and just imagine that all of these could be infused with the proper intellectual humility that comes from appreciating the amount unseen.

THE RATIONAL UNCONSCIOUS

ALISON GOPNIK

Psychologist, University of California–Berkeley; author, The Philosophical Baby: What Children's Minds Tell Us About Truth, Love, and the Meaning of Life

One of the greatest scientific insights of the twentieth century was that most psychological processes are not conscious. But the "unconscious" that made it into the popular imagination was Freud's irrational unconscious—the unconscious as a roiling, passionate id, barely held in check by conscious reason and reflection. This picture is still widespread, even though Freud has been largely discredited scientifically.

The "unconscious" that has actually led to the greatest scientific and technological advances might be called Turing's rational unconscious. If the version of the "unconscious" you see in movies like *Inception* were scientifically accurate, it would include phalanxes of nerds with slide rules instead of women in negligees wielding revolvers amid Daliesque landscapes. At least that might lead the audience to develop a more useful view of the mind—though probably not to buy more tickets.

Earlier thinkers like Locke and Hume anticipated many of the discoveries of psychological science but thought that the fundamental building blocks of the mind were conscious "ideas." Alan Turing, the father of the modern computer, began by thinking about the highly conscious and deliberate step-by-step calculations performed by human "computers" like the women decoding German ciphers at Bletchley Park. His first great insight was that the same processes could be instantiated in an entirely unconscious machine, with the same results. A machine could rationally decode the German ciphers using the same steps that the conscious "computers" went through. And the unconscious relay-and-vacuum-tube computers could get to the right answers in the same way the flesh-and-blood ones could.

Turing's second great insight was that we could understand much of the human mind and brain as an unconscious computer too. The women at Bletchley Park brilliantly performed conscious computations in their day jobs, but they were unconsciously performing equally powerful and accurate computations every time they spoke a word or looked across the room. Discovering the hidden messages about three-dimensional objects in the confusing mess of retinal images is just as difficult and important as discovering the hidden messages about submarines in the incomprehensible Nazi telegrams, and it turns out that the mind solves both mysteries in a similar way.

More recently, cognitive scientists have added the idea of probability into the mix, so that we can describe an unconscious mind, and design a computer, that can perform feats of inductive as well as deductive inference. Using this sort of probabilistic logic, a system can accurately learn about the world in a gradual, probabilistic way, raising the probability of some hypotheses and lowering that of others, and revising hypotheses in the light of new evidence. This work relies on a kind of reverse engineering. First, work out how any rational system could best infer the truth from the evidence it has. Often enough, you will find that the unconscious human mind does just that.

Some of the greatest advances in cognitive science have been the result of this strategy. But they have been largely invisible in popular culture, which has been understandably preoccupied with the sex and violence of much evolutionary psychology (like Freud, it makes for a better movie). Vision science studies how we are able to transform the chaos of stimulation at our retinas into a coherent and accurate perception of the outside world. It is, arguably, the most scientifically successful branch of both cognitive science and neuroscience. It takes off from the idea that our visual system is, entirely unconsciously, making rational inferences from retinal data to figure out what objects are like. Vision scientists began by figuring out the best way to solve the problem of vision and then discovered, in detail, just how the brain performs those computations.

The idea of the rational unconscious has also transformed our scientific understanding of creatures whose rationality has traditionally been denied, such as young children and animals. It should transform our everyday understanding, too. The Freudian picture identifies infants with that fantasizing, irrational unconscious; even in the classic Piagetian view, young children are profoundly illogical. But contemporary research shows the enormous gap between what young children say—and, presumably, what they experience—and their spectacularly accurate if unconscious feats of learning, induction, and reasoning. The rational unconscious gives us a way of understanding how babies can learn so much, when they seem to consciously understand so little.

Another way the rational unconscious could inform everyday thinking is by acting as a bridge between conscious experience and the few pounds of gray goo in our skulls. The gap between our experience and our brains is so great that people ping-pong between amazement and incredulity at every study showing that knowledge or love or goodness is "really in the brain" (though where else would it be?). There is important work linking the rational unconscious to both conscious experience and neurology.

Intuitively, we feel that we know our own minds—that our conscious experience is a direct reflection of what goes on underneath. But much of the most interesting work in social and cognitive psychology demonstrates the gulf between our rationally unconscious minds and our conscious experience. Our conscious understanding of probability, for example, is truly awful, in spite of the fact that we unconsciously make subtle probabilistic judgments all the time. The scientific study of consciousness has made us realize just how complex, unpredictable, and subtle the relation is between our minds and our experience.

At the same time, to be genuinely explanatory, neuroscience has to go beyond "the new phrenology" of simply locating psychological functions in particular brain regions. The rational unconscious lets us understand the how and why of the brain and not just the where. Again, vision science has led the way, with elegant empirical studies showing just how specific networks of neurons can act as computers rationally solving the problem of vision.

Of course, the rational unconscious has its limits. Visual illusions demonstrate that our brilliantly accurate visual system does sometimes get it wrong. Conscious reflection may be misleading sometimes, but it can also provide cognitive prostheses, the intellectual equivalent of glasses with corrective lenses, to help compensate for the limitations of the rational unconscious. The institutions of science do just that.

The greatest advantage of understanding the rational unconscious would be to demonstrate that rational discovery isn't a specialized abstruse privilege of the few we call scientists but is instead the evolutionary birthright of us all. Really tapping into our inner vision and inner child might not make us happier or better adjusted, but it might make us appreciate just how smart we really are.

WE ARE BLIND TO MUCH THAT SHAPES OUR MENTAL LIFE

ADAM ALTER

Psychologist; assistant professor of marketing, Stern School of Business; affiliated appointment, Department of Psychology, New York University

The human brain is an inconceivably complex tool. While we're focusing on the business of daily life, our brains are processing multitudes of information below the surface of conscious awareness; meanwhile this peripheral information subtly shapes our thoughts, feelings, and actions and crafts some of our most critical life outcomes. I'll illustrate with three brief examples from a larger set that comprise a forthcoming book I'll be publishing with Penguin Press:

1. Color

   Color is a ubiquitous feature of the environment, though we rarely notice colors unless they're particularly bright or deviate dramatically from our expectations. Nonetheless, they can shape a range of outcomes: A recent study conducted by University of Rochester psychologists Andrew Elliott and Daniela Niesta, for example, showed that men are slightly more attractive to women when they wear red shirts rather than shirts of another color. The same effect applies to women, who seem more attractive to men when their pictures are bordered in red. Red signals both romantic intent and dominance among lower-order species, and this same signal applies to men and women. This relationship between red and dominance explains findings by the evolutionary anthropologists Russell Hill and Robert Barton of the University of Durham (2005) that, "across a range of sports," contestants who wear red tend to outperform those wearing other colors. But red isn't always beneficial: We've come to associate it with errors and caution, so although it makes us more vigilant, it can also dampen our creativity (see, for instance, "Blue or Red? Exploring the Effect of Color on Cognitive Task Performances," by Ravi Mehta and Rui Zhu, in the February 27, 2009, issue of *Science*.)

All these effects have sound bases in biology and human psychology, but that doesn't make them any less remarkable or surprising.

2. Weather and ambient temperature

No one is surprised that the sunny warmth of summer makes people happy, but weather conditions and ambient temperature have other, more unexpected effects on our mental lives. Rainy weather makes us introspective and thoughtful, which in turn improves our memory [Forgas et al., 2009, *J. Exp. Soc. Psychol.*] In the Forgas study, people more accurately remembered the features of a store on rainy days than on sunny days. The stock market tends to rise on fine, sunny days, while cooler, rainy days prompt sluggishness and brief downturns (e.g., Hirshleifer & Shumway, 2001, *J. Finance;* Saunders, 1993, *Am. Econ. Rev.*). More surprising still is the relationship between changes in weather and suicide, depression, irritability, and various kinds of accidents—all of which are said to be responsive to changes in the electrical state of the atmosphere [Charry & Hawkinshire, 1981, *J. Pers. Soc. Psychol.*].

The association between warmth and human kindness is more than a metaphor; recent studies have shown that people find strangers more likable when they form their first impressions while holding a cup of hot coffee [Williams & Bargh, 2008, *Science*]. The warmth-kindness metaphor extends to social exclusion: People literally feel colder when they've been socially excluded.

3. Symbols and images
Urban landscapes are populated by thousands of symbols and images that unwittingly influence how we think and behave. My colleagues and I have found that self-identified Christians tend to behave more honestly when they're exposed to an image of the crucifix, even when they have no conscious memory of having seen it. A 1989 experiment conducted by psychologist Mark Baldwin of the Research Center for Group Dynamics at the University of Michigan, showed that Christians felt less virtuous after subliminal exposure to an image of Pope John Paul II, as they were reminded of the impossibly high standards of virtue demanded by religious authority.

On a brighter note, people tend to think more creatively when exposed to the Apple Computer logo [Fitzsimons et al., 2008, *J. Consumer Res.*], or when an incandescent light bulb is turned on [Slepian et al., 2010, *J. Exp. Soc. Psychol.*]; both the Apple logo and the illuminated light bulb are popularly associated with creativity, and deeply ingrained symbols, once activated, can shape how we think.

Similar associative logic suggests that national flags prompt unity; and, indeed, a sample of left- and right-wing Israelis were more accommodating of opposing political views when they were subliminally exposed to an image of the Israeli flag [Hassin et al., 2007, *Proc. Natl. Acad. Sci. USA*]. Likewise, a sample of Americans, when seated in front of a large U.S. flag, reported holding more positive attitudes toward Muslims [Butz et al., 2007, *Pers. Soc. Psychol. Bull.*].

These three cues—colors, weather conditions, and symbols and images—are joined by dozens of others that have a surprising ability to influence how we think, feel, behave, and decide. Once we understand what those cues are and how they shape our mental lives, we're better equipped to harness or discount them.

AN INSTINCT TO LEARN

W. TECUMSEH FITCH

Evolutionary biologist; professor of cognitive biology, University of Vienna; author, The Evolution of Language

One of the most pernicious misconceptions in cognitive science is the belief in a dichotomy between nature and nurture. Many psychologists, linguists, and social scientists, along with the popular press, continue to treat nature and nurture as combating ideologies rather than complementary perspectives. For such people, the idea that something is both "innate" and "learned," or both "biological" and "cultural," is an absurdity. Yet most biologists today recognize that understanding behavior requires that we understand the interaction between inborn cognitive processes (e.g., learning and memory) and individual experience. This is particularly true in human behavior, since the capacities for language and culture are some of the key adaptations of our species and involve irreducible elements of both biology and environment, of both nature and nurture.

The antidote to "nature versus nurture" thinking is to recognize the existence, and importance, of "instincts to learn." This phrase was introduced by Peter Marler, one of the fathers of birdsong research. A young songbird, while still in the nest, eagerly listens to adults of its own species sing. Months later, having fledged, it begins singing itself, and shapes its own initial sonic gropings to the template provided by those stored memories. During this period of "subsong," the bird gradually refines and perfects its own song, until by adulthood it is ready to defend a territory and attract mates with its own, perhaps unique, species-typical song.

Songbird vocal learning is the classic example of an instinct to learn. The songbird's drive to listen, and to sing, and to shape its song to that which it heard, is all instinctive. The bird needs no tutelage or feedback from its parents to go through these stages. Nonetheless, the actual song it sings is learned, passed culturally from generation to generation. Birds have local dialects, varying randomly from region to region. If the young bird hears no song, it will produce only an impoverished squawking, not a typical song.

Importantly, this capacity for vocal learning is true only of some birds, like songbirds and parrots. Other bird species—seagulls, chickens, owls—do not learn their vocalizations; rather, their calls develop reliably in the absence of any acoustic input. The calls of such birds are truly instinctive rather than learned. But for those birds capable of vocal learning, the song an adult bird sings is the result of a complex interplay between instinct (to listen, to rehearse, and to perfect) and learning (matching the songs of adults of its species).

It is interesting and perhaps surprising to realize that most mammals lack a capacity for complex vocal learning of this sort. Current research suggests that aside from humans, only marine mammals (whales, dolphins, seals), bats, and elephants have it. Among primates, humans appear to be the *only* species that can hear new sounds in the environment and then reproduce them. Our ability to do this seems to depend on a babbling stage during infancy, a period of vocal playfulness as instinctual as the young bird's subsong. During this stage, we appear to fine-tune our vocal control so that, as children, we can hear and reproduce the words and phrases of our adult caregivers.

So is human language an instinct or learned? The question, presupposing a dichotomy, is intrinsically misleading. Every word that any human speaks, in any of our species' six thousand languages, has been learned. And yet the *capacity* for learning that language is a human instinct, something that every normal human child is born with and that no chimpanzee or gorilla possesses.

The instinct to learn language is, indeed, innate (meaning simply that it reliably develops in our species), even though every language is learned. As Darwin put it in *The Descent of Man*, "language is an art, like brewing or baking; but . . . certainly is not a true instinct, for every language has to be learnt. It differs, however, widely from all ordinary arts, for man has an instinctive tendency to speak, as we see in the babble of our young children; whilst no child has an instinctive tendency to brew, bake, or write."

And what of culture? For many, human culture seems the very antithesis of "instinct." And yet it must be true that language plays a key role in every human culture. Language is the primary medium for the passing on of historically accumulated knowledge, tastes, biases, and styles that makes each of our human tribes and nations its own unique and precious entity. And if human language is best conceived of as an instinct to learn, why not culture itself?

The past decade has seen a remarkable unveiling of our human genetic and neural makeup, and the coming decade promises even more remarkable breakthroughs. Each of us 6 billion humans is genetically unique (with the fascinating exception of identical twins). For each of us, our unique genetic makeup influences, but does not determine, what we are.

If we are to grapple earnestly and effectively with the reality of human biology and genetics, we will need to jettison outmoded dichotomies like the traditional distinction between nature and nurture. In their place, we will need to embrace the reality of the many instincts to learn (language, music, dance, culture) that make us human.

I conclude that the dichotomy-denying phrase "instinct to learn" deserves a place in the cognitive toolkit of everybody who hopes, in the coming age of individual genomes, to understand human culture and human nature in the context of human biology. Human language, and human culture, are not instincts—but they *are* instincts to learn.

THINK BOTTOM UP, NOT TOP DOWN

MICHAEL SHERMER

Publisher of Skeptic *magazine; adjunct professor, Claremont Graduate University; author,* The Believing Brain: From Ghosts and Gods to Politics and Conspiracies—How We Construct Beliefs and Reinforce Them as Truths

One of the most general shorthand abstractions that, if adopted, would improve the cognitive toolkit of humanity is to *think bottom up, not top down*. Almost everything important that happens in both nature and society happens from the bottom up, not the top down. Water is a bottom-up, self-organized emergent property of hydrogen and oxygen. Life is a bottom-up, self-organized emergent property of organic molecules that coalesced into protein chains through nothing more than the input of energy into the system of Earth's early environment. The complex eukaryotic cells of which we are made are themselves the product of much simpler prokaryotic cells that merged together from the bottom up, in a process of symbiosis that happens naturally when genomes are merged between two organisms. Evolution itself is a bottom-up process of organisms just trying to make a living and get their genes into the next generation; out of that simple process emerges the diverse array of complex life we see today.

Analogously, an economy is a self-organized bottom-up emergent process of people just trying to make a living and get their genes into the next generation, and out of that simple process emerges the diverse array of products and services available to us today. Likewise, democracy is a bottom-up emergent political system specifically designed to displace top-down kingdoms, theocracies, and dictatorships. Economic and political systems are the result of human action, not human design.

Most people, however, see the world from the top down instead of the bottom up. The reason is that our brains evolved to find design in the world, and our experience with designed objects is that they have a designer (us), whom we consider to be intelligent. So most people intuitively sense that anything in nature that looks designed must be so from the top down, not the bottom up. Bottom-up reasoning is counterintuitive. This is why so many people believe that life was designed from the top down, and why so many think that economies must be designed and that countries should be ruled from the top down.

One way to get people to adopt the bottom-up shorthand abstraction as a cognitive tool is to find examples that we *know* evolved from the bottom up and were not designed from the top down. Language is an example. No one designed English to look and sound like it does today (in which teenagers use the word "like" in every sentence). From Chaucer's time forward, our language has evolved from the bottom up by native speakers adopting their own nuanced styles to fit their unique lives and cultures. Likewise, the history of knowledge production has been one long trajectory from top down to bottom up. From ancient priests and medieval scholars to academic professors and university publishers, the democratization of knowledge has struggled alongside the democratization of societies to free itself from the bondage of top-down control. Compare the magisterial multivolume encyclopedias of centuries past that held sway as the final authority for reliable knowledge, now displaced by individual encyclopedists employing wiki tools and making everyone his own expert.

Which is why the Internet is the ultimate bottom-up self-organized emergent property of millions of computer users exchanging data across servers, and although there are some top-down controls involved—just as there are some in mostly bottom-up economic and political systems—the strength of digital freedom derives from the fact that no one is in charge. For the past five hundred years, humanity has gradually but ineluctably transitioned from top-down to bottom-up systems, for the simple reason that both information and people want to be free.

FIXED-ACTION PATTERNS

IRENE PEPPERBERG
Research associate, Harvard University; adjunct associate professor in psychology, Brandeis University; author, Alex & Me: How a Scientist and a Parrot Discovered a Hidden World of Animal Intelligence—and Formed a Deep Bond in the Process

The concept comes from early ethologists, scientists such as Oskar Heinroth and Konrad Lorenz, who defined it as an instinctive response—usually a series of predictable behavior patterns—that would occur reliably in the presence of a specific bit of input, often called a "releaser." Fixed-action patterns, as such patterns were known, were thought to be devoid of cognitive processing. As it turned out, fixed-action patterns were not nearly as fixed as the ethologists believed, but the concept has remained as part of the historical literature—as a way of scientifically describing what in the vernacular might be called "knee-jerk" responses. The concept of a fixed-action pattern, despite its simplicity, may prove valuable as a metaphorical means to study and change human behavior.

If we look into the literature on fixed-action patterns, we see that many such instinctive responses were actually learned, based on the most elementary of signals. For example, the newly hatched herring gull chick's supposed fixed-action pattern of hitting the red spot on its parent's beak for food was far more complex. Ornithologist and ethologist Jack P. Hailman demonstrated that what was innate was only a tendency to peck at an oscillating object in the field of view. The ability to target the beak, and the red spot on the beak, though a pattern that developed steadily and fairly quickly, was acquired experientially. Clearly, certain sensitivities must be innate, but the specifics of their development into various behavioral acts likely depend on how the organism interacts with its surroundings and what feedback it receives. The system need not, especially for humans, be simply a matter of conditioning response R to stimulus S but rather of evaluating as much input as possible.

The relevance is that if we wish to understand why, as humans, we often act in certain predictable ways (and particularly if there is a desire or need to change these behavioral responses), we can remember our animal heritage and look for the possible releasers that seem to stimulate our fixed-action patterns. Might the fixed-action pattern actually be a response learned over time, initially with respect to something even more basic than we expect? The consequences could affect several aspects of our lives, from our social interactions to the quick decision-making processes in our professional roles. Given an understanding of our fixed-action pattern, and those of the individuals with whom we interact, we—as humans with cognitive processing powers—could begin to rethink our behavior patterns.

POWERS OF 10

TERRENCE SEJNOWSKI
Computational neuroscientist; Francis Crick Professor, the Salk Institute; coauthor (with Patricia Churchland), The Computational Brain

An important part of my scientific toolkit is how to think about things in the world over a wide range of magnitudes and time scales. This involves, first, understanding powers of ten; second, visualizing data over a wide range of magnitudes on graphs using logarithmic scales; and third, appreciating the meaning of magnitude scales, such as the decibel (dB) scale for the loudness of sounds and the Richter scale for the strength of earthquakes.

This toolkit ought to be a part of everyone's thinking, but sadly I have found that even well-educated nonscientists are flummoxed by log scales and can only vaguely grasp the difference between an earthquake of 6 on the Richter scale and one of 8 (a thousand times more energy released). Thinking in powers of 10 is such a basic skill that it ought to be taught along with integers in elementary school.

Scaling laws are found throughout nature. Galileo in 1638 pointed out that large animals have disproportionately thicker leg bones than small animals to support the weight of the animal. The heavier the animal, the stouter their legs need to be. This leads to a prediction that the thickness of the leg bone should scale with the 3/2 power of the length of the bone.

Another interesting scaling law is that between the volume of the brain's cortical white matter, corresponding to the long-distance wires between cortical areas, and the gray matter, where the computing takes place. For mammals ranging over 5 orders of magnitude in weight—from a pygmy shrew to an elephant—the white matter scales as the 5/4 power of the gray matter. This means that the bigger the brain, the disproportionately larger the fraction of the volume taken up by cortical wiring used for communication compared to the computing machinery.

I am concerned that students I teach have lost the art of estimating with powers of 10. When I was a student, I used a slide rule to compute, but students now use calculators. A slide rule lets you carry out a long series of multiplications and divisions by adding and subtracting the logs of numbers; but at the end you need to figure out the powers of 10 by making a rough estimate. A calculator keeps track of this for you, but if you make a mistake in keying in a number, you can be off by 10 orders of magnitude, which happens to students who don't have a feeling for orders of magnitude.

A final reason why familiarity with powers of 10 would improve everyone's cognitive toolkit is that it helps us comprehend our life and the world in which we live:

How many seconds are there in a lifetime? 10^9 sec

A second is an arbitrary time unit, but one that is based on our experience. Our visual system is bombarded by snapshots at a rate of around three per second, caused by rapid eye movements called saccades. Athletes often win or lose a race by a fraction of a second. If you earned a dollar for every second in your life, you would be a billionaire. However, a second can feel like a minute in front of an audience, and a quiet weekend can disappear in a flash. When I was a child, a summer seemed to last forever, but now the summer is over almost before it begins. William James speculated that subjective time was measured in novel experiences, which become rarer as you get older. Perhaps life is lived on a logarithmic time scale, compressed toward the end.

What is the GDP of the world? $\$10^{14}$

A billion dollars was once worth a lot, but there is now a long list of multibillionaires. The U.S. government recently stimulated the world economy by loaning several trillion dollars to banks. It is difficult to grasp how much a trillion dollars ($\$10^{12}$) represents, but several clever videos on YouTube (key words: trillion dollars) illustrate this with physical comparisons (a giant pile of $100 bills) and what you can buy with it (ten years of U.S. response to 9/11). When you start thinking about the world economy, the trillions of dollars add up. A trillion here, a trillion there, pretty soon you're talking about real money. But so far there aren't any trillionaires.

How many synapses are there in the brain? 10^{15}

Two neurons can communicate with each other at a synapse, which is the computational unit in the brain. The typical cortical synapse is less than a micron in diameter (10^{-6} meter), near the resolution limit of the light microscope. If the economy of the world is a stretch for us to contemplate, thinking about all the synapses in your head is mind-boggling. If I had a dollar for every synapse in your brain, I could support the current economy of the world for ten years. Cortical neurons on average fire once a second, which implies a bandwidth of around 10^{15} bits per second, greater than the total bandwidth of the Internet backbone.

How many seconds will the sun shine? 10^{17} sec

Our sun has shined for billions of years and will continue to shine for billions more. The universe seems to be standing still during our lifetime, but on longer time scales the universe is filled with events of enormous violence. The spatial scales are also immense. Our space-time trajectory is a very tiny part of the universe, but we can at least attach powers of 10 to it and put it into perspective.

LIFE CODE

JUAN ENRIQUEZ
Managing director, Excel Venture Management; author, As the Future Catches You: How Genomics & Other Forces Are Changing Your Life, Work, Health & Wealth; *coauthor (with Steve Gullans),* Homo Evolutis: Please Meet the Next Human Species

Everyone is familiar with digital code, the shorthand IT. Soon all may be discoursing about life code.

It took a while to learn how to read life code; Mendel's initial cookbook was largely ignored. Darwin knew but refused for decades to publish such controversial material. Even DNA, a term that now lies within every cheesy PR description of a firm, on jeans, and in pop psych books, was largely ignored after its 1953 discovery. For close to a decade, very few cited Watson and Crick. They were not even nominated, by anyone, for a Nobel until after 1960, despite their discovery of how life code is written.

First ignorance, then controversy, continued dogging life code as humanity moved from reading it to copying it. Tadpoles were cloned in 1952, but few focused on that process until 1997, when the announcement of the cloning of Dolly the sheep begat wonder, consternation, and fear. Much the same occurred with in-vitro fertilization and Louise Brown, a breakthrough that got the Nobel in 2010, a mere thirty-two years after the first birth. Copying genes of dozens of species, leading to hundreds of thousands of almost identical animals, is now commonplace. The latest controversy is no longer "How do we deal with clones?" but "Should we eat them?"

Much has occurred as we learned to read and copy life code, but there is still little understanding of recent developments. Yet it is this third stage—writing and rewriting life code—that is by far the most important and profound.

Few realize, so far, that life code is spreading across industries, economies, countries, and cultures. As we begin to rewrite existing life, strange things evolve. Bacteria can be programmed to solve Sudoku puzzles. Viruses begin to create electronic circuits. As we write life from scratch, J. Craig Venter, Hamilton Smith, et al., partner with Exxon to try to change the world's energy markets. Designer genes introduced by retroviruses, organs built from scratch, the first synthetic cells—these are further examples of massive change.

We see more and more products derived from life code changing fields as diverse as energy, textiles, chemicals, IT, vaccines, medicines, space exploration, agriculture, fashion, finance, and real estate. And gradually, "life code," a concept with only 559 Google hits in 2000 and fewer than 50,000 in 2009, becomes a part of everyday public discourse.

Much has occurred over the past decades with digital code, leading to the likes of Digital, Lotus, HP, IBM, Microsoft, Amazon, Google, and Facebook. Many of the Fortune 500 companies within the next decade will be based on the understanding and application of life code.

But this is just the beginning. The real change will become apparent as we rewrite life code to morph the human species. We are already transitioning from a humanoid that is shaped by and shapes its environment into a *Homo evolutis*, a species that directly and deliberately designs and shapes its own evolution and that of other species.

CONSTRAINT SATISFACTION

STEPHEN M. KOSSLYN

Director, Center for Advanced Study in the Behavioral Sciences, Stanford University; author, Image and Mind

The concept of constraint satisfaction is crucial for understanding and improving human reasoning and decision making. A "constraint" is a condition that must be taken into account when solving a problem or making a decision, and "constraint satisfaction" is the process of meeting the relevant constraints. The key idea is that often there are only a few ways to satisfy a full set of constraints simultaneously.

For example, when moving into a new house, my wife and I had to decide how to arrange the furniture in the bedroom. We had an old headboard, which was so rickety that it had to be leaned against a wall. This requirement was a constraint on the positioning of the headboard. The other pieces of furniture also had requirements (constraints) on where they could be placed. Specifically, we had two small end tables that had to be next to either side of the headboard; a chair that needed to be somewhere in the room; a reading lamp that needed to be next to the chair; and an old sofa that was missing one of its rear legs and hence rested on a couple of books—and we wanted to position it so that people couldn't see the books. Here was the remarkable fact about our exercises in interior design: Virtually always, as soon as we selected the wall for the headboard, *bang!* The entire configuration of the room was determined. There was only one other wall large enough for the sofa, which in turn left only one space for the chair and lamp.

In general, the more constraints, the fewer the possible ways of satisfying them simultaneously. And this is especially the case when there are many "strong" constraints. A strong constraint is like the positioning of the end tables: There are very few ways to satisfy it. In contrast, a "weak" constraint, such as the location of the headboard, can be satisfied in many ways. (Many positions along different walls would work.)

What happens when some constraints are incompatible with others? For instance, say that you live far from a gas station and so you want to buy an electric automobile—but you don't have enough money to buy one. Not all constraints are equal in importance, and as long as the most important ones are satisfied "well enough," you may have reached a satisfactory solution. In this case, although an optimal solution to your transportation needs might be an electric car, a hybrid that gets excellent gas mileage might be good enough.

In addition, once you begin the constraint-satisfaction process, you can make it more effective by seeking out additional constraints. For example, when you're deciding what car to buy, you might start with the constraints of (a) your budget and (b) your desire to avoid going to a gas station. You then might consider the size of car needed for your purposes, length of the warranty, and styling. You may be willing to make tradeoffs—for example, by satisfying some constraints very well (such as mileage) but just barely satisfying others (e.g., styling). Even so, the mere fact of including additional constraints could be the deciding factor. Constraint satisfaction is pervasive. For example:

- This is how detectives—from Sherlock Holmes to the Mentalist—crack their cases, treating each clue as a constraint and looking for a solution that satisfies them all.
- This is what dating services strive to do—find the clients' constraints, identify which constraints are most important to

him or her, and then see which of the available candidates best satisfies the constraints.

- This is what you go through when finding a new place to live, weighing the relative importance of constraints such as the size, price, location, and type of neighborhood.
- And this is what you do when you get dressed in the morning: You choose clothes that "go with each other" (both in color and style).

Constraint satisfaction is pervasive in part because it does not require "perfect" solutions. It's up to you to decide what the most important constraints are and just how many of the constraints in general must be satisfied (and how well). Moreover, constraint satisfaction need not be linear: You can appreciate the entire set of constraints at the same time, throwing them into your mental stewpot and letting them simmer. And this process need not be conscious. "Mulling it over" seems to consist of engaging in all-but-unconscious constraint satisfaction.

Finally, much creativity emerges from constraint satisfaction. Many new recipes have been created when chefs discovered that only certain ingredients were available—and they thus were either forced to substitute for those missing or come up with a new dish. Creativity can also emerge when you decide to change, exclude, or add a constraint. Einstein had one of his major breakthroughs when he realized that time need not pass at a constant rate. Perhaps paradoxically, adding constraints can actually enhance creativity—if a task is too open or unstructured, it may be so unconstrained that it's difficult to devise any solution.

CYCLES

DANIEL C. DENNETT
Philosopher; professor and codirector, Center for Cognitive Studies, Tufts University; author, Breaking the Spell: Religion as a Natural Phenomenon

Everybody knows about the familiar large-scale cycles of nature: Day follows night follows day, summer-fall-winter-spring-summer-fall-winter-spring, the water cycle of evaporation and precipitation that refills our lakes, scours our rivers, and restores the water supply of every living thing on the planet. But not everybody appreciates how cycles—every spatial and temporal scale from the atomic to the astronomic—are quite literally the hidden spinning motors that power all the wonderful phenomena of nature.

Nikolaus Otto built and sold the first internal-combustion gasoline engine in 1861, and Rudolf Diesel built his engine in 1897, two brilliant inventions that changed the world. Each exploits a cycle, the four-stroke Otto cycle or the two-stroke Diesel cycle, that accomplishes some work and then restores the system to the original position so that it is ready to accomplish some more work. The details of these cycles are ingenious, and they have been discovered and optimized by an R & D cycle of invention that is several centuries old. An even more elegant, micro-miniaturized engine is the Krebs cycle, discovered in 1937 by Hans Krebs but invented over millions of years of evolution at the dawn of life. It is the eight-stroke chemical reaction that turns fuel into energy in the process of metabolism that is essential to all life, from bacteria to redwoods.

Biochemical cycles like the Krebs cycle are responsible for all the motion, growth, self-repair, and reproduction in the living world, wheels within wheels within wheels, a clockwork with trillions of moving parts, and each clock has to be rewound, restored to step one so that it can do its duty again. All of these have been optimized by the grand Darwinian cycle of reproduction, generation after generation, picking up fortuitous improvements over the eons.

At a completely different scale, our ancestors discovered the efficacy of cycles in one of the great advances of human prehistory: the role of repetition in manufacture. Take a stick and rub it with a stone and almost nothing happens—a few scratches are the only visible sign of change. Rub it a hundred times and there is still nothing much to see. But rub it just so, for a few thousand times, and you can turn it into an uncannily straight arrow shaft. By the accumulation of imperceptible increments, the cyclical process creates something altogether new. The foresight and self-control required for such projects was itself a novelty, a vast improvement over the repetitive but largely instinctual and mindless building and shaping processes of other animals. And that novelty was, of course, itself a product of the Darwinian cycle, enhanced eventually by the swifter cycle of cultural evolution, in which the reproduction of the technique wasn't passed on to offspring through the genes but transmitted among non-kin conspecifics who picked up the trick of imitation.

The first ancestor who polished a stone into a handsomely symmetrical hand axe must have looked pretty stupid in the process. There he sat, rubbing away for hours on end, to no apparent effect. But hidden in the interstices of all the mindless repetition was a process of gradual refinement that was well nigh invisible to the naked eye designed by evolution to detect changes occurring at a much faster tempo. The same appearance of futility has occasionally misled sophisticated biologists. In his elegant book *Wetware*, the molecular and cell biologist Dennis Bray describes cycles in the nervous system:

In a typical signaling pathway, proteins are continually being modified and demodified. Kinases and phosphatases work ceaselessly like ants in a nest, adding phosphate groups to proteins and removing them again. It seems a pointless exercise, especially when you consider that each cycle of addition and removal costs the cell one molecule of ATP—one unit of precious energy. Indeed, cyclic reactions of this kind were initially labeled "futile." But the adjective is misleading. The addition of phosphate groups to proteins is the single most common reaction in cells and underpins a large proportion of the computations they perform. Far from being futile, this cyclic reaction provides the cell with an essential resource: a flexible and rapidly tunable device.

The word "computations" is aptly chosen, for it turns out that all the "magic" of cognition depends, just as life itself does, on cycles within cycles of recurrent, re-entrant, reflexive information-transformation processes, from the biochemical scale within the neuron to the whole brain sleep cycle, waves of cerebral activity and recovery revealed by EEGs. Computer programmers have been exploring the space of possible computations for less than a century, but their harvest of invention and discovery so far includes millions of loops within loops within loops. The secret ingredient of improvement is always the same: practice, practice, practice.

It is useful to remember that *Darwinian* evolution is just one kind of accumulative, refining cycle. There are plenty of others. The problem of the origin of life can be made to look insoluble ("irreducibly complex") if one argues, as intelligent-design advocates have done, that "since evolution by natural selection depends on reproduction," there cannot be a Darwinian solution to the problem of how the first living, reproducing thing came to exist.

It was surely breathtakingly complicated, beautifully designed—must have been a miracle.

If we lapse into thinking of the prebiotic, pre-reproductive world as a sort of featureless chaos of chemicals (like the scattered parts of the notorious jetliner assembled by a windstorm), the problem does look daunting and worse, but if we remind ourselves that the key process in evolution is cyclical repetition (of which genetic replication is just one highly refined and optimized instance), we can begin to see our way to turning the mystery into a puzzle: How did all those seasonal cycles, water cycles, geological cycles, and chemical cycles, spinning for millions of years, gradually accumulate the preconditions for giving birth to the biological cycles? Probably the first thousand "tries" were futile, near misses. But as the wonderfully sensual song by Gershwin and DeSylva reminds us, see what happens if you "Do it again" (and again, and again).

A good rule of thumb, then, when confronting the apparent magic of the world of life and mind is: Look for the cycles that are doing all the hard work.

KEYSTONE CONSUMERS

JENNIFER JACQUET
Postdoctoral researcher in environmental economics, University of British Columbia

When it comes to common resources, a failure to cooperate is a failure to control consumption. In Garrett Hardin's classic tragedy, everyone overconsumes and equally contributes to the detriment of the commons. But a relative few can also ruin a resource for the rest of us.

Biologists are familiar with the term "keystone species," coined in 1969 after Robert Paine's intertidal exclusion experiments. Paine found that by removing the few five-limbed carnivores—the purple sea star, *Pisaster ochraceus*—from the seashore, he could cause an overabundance of its prey, mussels, and a sharp decline in diversity. Without sea stars, mussels outcompeted sponges. No sponges, no nudibranchs. Anemones were also starved out, because they eat what the sea stars dislodge. *Pisaster* was the keystone that kept the intertidal community together. Without it, there were only mussels, mussels, mussels. The term "keystone species," inspired by the purple sea star, refers to a species that has a disproportionate effect relative to its abundance.

In human ecology, I imagine diseases and parasites play a similar role to *Pisaster* in Paine's experiment. Remove disease (and increase food) and *Homo sapiens* takes over. Humans inevitably restructure their environment. But not all human beings consume equally. While a keystone species refers to a specific species that structures an ecosystem, I consider keystone consumers to be a specific group of humans that structures a market for a particular resource. Intense demand by a few individuals can bring flora and fauna to the brink.

There are keystone consumers in the markets for caviar, slipper orchids, tiger penises, plutonium, pet primates, diamonds, antibiotics, Hummers, and sea horses. Niche markets for frog legs in pockets of the United States, Europe, and Asia are depleting frog populations in Indonesia, Ecuador, and Brazil. Seafood lovers in high-end restaurants are causing stocks of long-lived fish species like orange roughy or toothfish in Antarctica to crash. The desire of wealthy Chinese consumers for shark-fin soup has led to the collapse of several shark species.

One in every four species of mammals (1,141 of the 5,487 mammalian species on Earth) is threatened with extinction. At least seventy-six mammalian species have become extinct since the sixteenth century—many, like the Tasmanian tiger, the great auk, and Steller's sea cow, because of hunting by a relatively small group. It is possible for a small minority of humans to precipitate the disappearance of an entire species.

The consumption of nonliving resources is also imbalanced. The 15 percent of the world's population that lives in North America, Western Europe, Japan, and Australia consumes thirty-two times more resources, like fossil fuels and metals, and produces thirty-two times more pollution than the developing world, where the remaining 85 percent of humans live. City dwellers consume more than people living in the countryside. A recent study determined that the ecological footprint for an average resident of Vancouver, British Columbia, was thirteen times higher than that of his suburban/rural counterpart.

Developed nations, urbanites, ivory collectors: The keystone consumer depends on the resource in question. In the case of water, agriculture accounts for 80 percent of use in the United States;

large-scale farms are the keystone consumers. So why do many conservation efforts focus on households rather than on water efficiency on farms? The keystone-consumer concept helps focus conservation efforts where returns on investments are highest.

Like keystone species, keystone consumers also have a disproportionate impact relative to their abundance. Biologists identify keystone species as conservation priorities because their disappearance could cause the loss of many other species. In the marketplace, keystone consumers should be priorities because their disappearance could lead to the recovery of the resource. Humans should protect keystone species and curb keystone consumption. The lives of others depend on it.

CUMULATIVE ERROR

JARON LANIER
Musician, computer scientist; virtual reality pioneer; author,
You Are Not a Gadget: A Manifesto

It is the stuff of children's games. In the game of "Telephone," a secret message is whispered from child to child until it is announced out loud by the final recipient. To the delight of all, the message is typically transformed into something new and bizarre, no matter the sincerity and care given to each retelling.

Humor seems to be the brain's way of motivating itself—through pleasure—to notice disparities and cleavages in its sense of the world. In the telephone game, we find glee in the violation of expectation; what we think should be fixed turns out to be fluid.

When brains get something wrong commonly enough that noticing the failure becomes the fulcrum of a simple child's game, then you know there's a hitch in human cognition worth worrying about. Somehow, we expect information to be Platonic and faithful to its origin, no matter what history might have corrupted it.

The illusion of Platonic information is confounding because it can easily defeat our natural skeptical impulses. If a child in the sequence sniffs that the message seems too weird to be authentic, she can compare notes most easily with the children closest to her, who received the message just before she did. She might discover some small variation, but mostly the information will appear to be confirmed, and she will find an apparent verification of a falsity.

Another delightful pastime is overtransforming an information artifact through digital algorithms—useful, if used sparingly—until it turns into something quite strange. For instance, you can use one of the online machine-translation services to translate a phrase through a ring of languages back to the original and see what you get. The sentence "The edge of knowledge motivates intriguing online discussions" transforms into "Online discussions in order to stimulate an attractive national knowledge" in four steps on Google's current translator (English→German→Hebrew→Simplified Chinese→English). We find this sort of thing funny, just like children playing "Telephone"—as well we should, because it sparks our recollection that our brains have unrealistic expectations of information transformation.

While information technology can reveal truths, it can also create stronger illusions than we are used to. For instance, sensors all over the world, connected through cloud computing, can reveal urgent patterns of change in climate data. But endless chains of online retelling also create an illusion for masses of people that the original data is a hoax.

The illusion of Platonic information plagues finance. Financial instruments are becoming multilevel derivatives of the real actions on the ground, which finance is ultimately supposed to motivate and optimize. The reason to finance the buying of a house ought to be at least in part to get the house bought. But an empire of specialists and giant growths of cloud computers showed, in the run-up to the Great Recession, that it is possible for sufficiently complex financial instruments to become completely disconnected from their ultimate purpose.

In the case of complex financial instruments, the role of each child in the telephone game corresponds not to a horizontal series of stations that relay a message but to a vertical series of transformations that are no more reliable. Transactions are stacked on top of each other. Each transaction is based on a formula that transforms the data of the transactions beneath it on the stack. A transaction might be based on the possibility that a prediction of a prediction will have been improperly predicted.

The illusion of Platonic information reappears as a belief that a higher-level representation must always be better. Each time a transaction is gauged to an assessment of the risk of another transaction, however, even if it is in a vertical structure, at least a little bit of error and artifact is injected. By the time a few layers have been compounded, the information becomes bizarre.

Unfortunately, the feedback loop that determines whether a transaction is a success or not is based only on its interactions with its immediate neighbors in the phantasmagorical abstract playground of finance. So a transaction can make money based on how it interacted with the other transactions it referenced directly, while having no relationship to the real events on the ground that all the transactions are ultimately rooted in. This is just like the child trying to figure out whether or not a message has been corrupted by talking only to her neighbors.

In principle, the Internet can make it possible to connect people directly to information sources, to avoid the illusions of the game of Telephone. Indeed, this happens. Millions of people had, for example, a remarkable direct experience of the Mars rovers.

The economy of the Internet as it has evolved incentivizes aggregators, however. Thus we all take seats in a new game of Telephone, in which you tell the blogger, who tells the aggregator of blogs, who tells the social network, who tells the advertiser, who tells the political action committee, and so on. Each station along the way finds that it is making sense, because it has the limited scope of the skeptical girl in the circle, and yet the whole system becomes infused with a degree of nonsense.

A joke isn't funny anymore if it's repeated too much. It is urgent for the cognitive fallacy of Platonic information to be universally acknowledged and for information systems to be designed to reduce cumulative error.

CULTURAL ATTRACTORS

DAN SPERBER
Philosophy and cognitive science professor, Central European University, Budapest

In 1967, Richard Dawkins introduced the idea of a meme, a unit of cultural transmission capable of replicating itself and undergoing Darwinian selection. "Meme" has become a remarkably successful addition to everybody's cognitive toolkit. I want to suggest that the concept of a meme should be, if not replaced by, at least supplemented with that of a *cultural attractor.*

The very success of the word "meme" is, or so it seems, an illustration of the idea of a meme: The word has now been used billions of time. But is the idea of a meme being replicated whenever the word is being used? Well, no. Not only do "memeticists" have many quite different definitions of "meme," but also, and more important, most users of the term have no clear idea of what a meme might be. The term is used with a vague meaning, relevant in the circumstances. These meanings overlap, but they are not replications of one another. The idea of a meme, as opposed to the word "meme," may not be such a good example of a meme after all!

The case of the meme idea illustrates a general puzzle. Cultures do contain items—ideas, norms, tales, recipes, dances, rituals, tools, practices, and so on—that are produced again and again. These items remain self-similar over social space and time. In spite of variations, an Irish stew is an Irish stew, Little Red Riding Hood is Little Red Riding Hood, and a samba is a samba. The obvious way to explain this stability at the macro level of the culture is, or so it seems, to assume fidelity at the micro level of interindividual transmission. Little Red Riding Hood must have been replicated faithfully enough most of the time for the tale to have remained self-similar over centuries of oral transmission, or else the story would have drifted in all kinds of ways and the tale itself would have vanished, like water in the sand. Macro stability implies micro fidelity. Right? Well, no. When we study micro processes of transmission—leaving aside those that use techniques of strict replication, such as printing or Internet forwarding—what we observe is a mix of *preservation* of the model and *construction* of a version that suits the capacities and interests of the transmitter. From one version to the next the changes may be small, but when they occur at the population scale, their cumulative effect should compromise the stability of cultural items. But—and here lies the puzzle—they don't. What, if not fidelity, explains stability?

Well, bits of culture—memes, if you want to dilute the notion and call them that—remain self-similar not because they are replicated again and again but because variations that occur at almost every turn in their repeated transmission, rather than resulting in "random walks" drifting away in all directions from an initial model, tend to gravitate around cultural attractors. Ending Little Red Riding Hood when the wolf eats the child would make for a simpler story to remember, but a Happy Ending is too powerful a cultural attractor. If someone had heard only the story ending with the wolf's meal, my guess is that either she would not have retold it at all (and that is selection) or she would have modified it by reconstructing a happy ending (and that is attraction). Little Red Riding Hood has remained culturally stable not because it has been faithfully replicated all along but because the variations present in all its versions have tended to cancel one another out.

Why should there be cultural attractors at all? Because there are in our minds, our bodies, and our environment biasing factors that affect the way we interpret and re-produce ideas and behaviors. (I write "re-produce" with a hyphen because, more often than not, we produce a new token of the same type without reproducing in the usual sense of copying some previous token.) When these biasing factors are shared in a population, cultural attractors emerge. Here are a few rudimentary examples.

Rounded numbers are cultural attractors: They are easier to remember and provide better symbols for magnitudes. So we celebrate twentieth wedding anniversaries, hundredth issues of journals, the millionth copy sold of a record, and so on. This, in turn, creates a special cultural attractor for prices, just below rounded numbers—$9.99 or $9,990 are likely price tags—so as to avoid the evocation of a higher magnitude.

In the diffusion of techniques and artifacts, efficiency is a powerful cultural attractor. Paleolithic hunters learning from their elders how to manufacture and use bows and arrows were aiming not so much at copying the elders as at becoming themselves as good as possible at shooting arrows. Much more than faithful replication, this attraction of efficiency when there aren't that many ways of being efficient explains the cultural stability (and also the historical transformations) of various technical traditions.

In principle, there should be no limit to the diversity of supernatural beings that humans can imagine. However, as the anthropologist Pascal Boyer has argued, only a limited repertoire of such beings is exploited in human religions. Its members—ghosts, gods, ancestor spirits, dragons, and so on—have in common two features:

1. They each violate some major intuitive expectations about living beings: the expectation of mortality, of belonging to one and only one species, of being limited in one's access to information, and so on.

2. They satisfy all other intuitive expectations and are therefore, in spite of their supernaturalness, rather predictable.

Why should this be so? Because being "minimally counterintuitive" (Boyer's phrase) makes for "relevant mysteries" (my phrase) and is a cultural attractor. Imaginary beings that are either less or more counterintuitive than that are forgotten or are transformed in the direction of this attractor.

And what is the attractor around which the "meme" meme gravitates? The meme idea—or rather a constellation of trivialized versions of it—has become an extraordinarily successful bit of contemporary culture not because it has been faithfully replicated again and again but because our conversation often does revolve (and here is the cultural attractor) around remarkably successful bits of culture that, in the time of mass media and the Internet, pop up more and more frequently and are indeed quite relevant to our understanding of the world. They attract our attention even when—or, possibly, especially when—we don't understand that well what they are and how they come about. The meaning of "meme" has drifted from Dawkins's precise scientific idea to a means to refer to these striking and puzzling objects.

This was my answer. Let me end by posing a question (which time will answer): Is the idea of a cultural attractor itself close enough to a cultural attractor for a version of it to become in turn a "meme"?

SCALE ANALYSIS

GIULIO BOCCALETTI
Physicist; atmospheric and oceanic scientist; expert associate principal, McKinsey & Company

There is a well-known saying: Dividing the universe into things that are linear and those that are nonlinear is very much like dividing the universe into things that are bananas and things that are not. Many things are not bananas.

Nonlinearity is a hallmark of the real world. It occurs any time that outputs of a system cannot be expressed in terms of a sum of inputs, each multiplied by a simple constant—a rare occurrence in the grand scheme of things. Nonlinearity does not necessarily imply complexity, just as linearity does not exclude it, but most real systems do exhibit some nonlinear feature that results in complex behavior. Some, like the turbulent stream from a water tap, hide deep nonlinearity under domestic simplicity, while others—weather, for example—are evidently nonlinear to the most distracted of observers. Nonlinear complex dynamics are around us: Unpredictable variability, tipping points, sudden changes in behavior, hysteresis—all are frequent symptoms of a nonlinear world.

Nonlinear complexity has also the unfortunate characteristic of being difficult to manage, high-speed computing notwithstanding, because it tends to lack the generality of linear solutions. As a result, we have a tendency to view the world in terms of linear models—for much the same reason that looking for lost keys under a lamppost might make sense because that's where the light is. Understanding seems to require simplification, one in which complexity is reduced wherever possible and only the most material parts of the problem are preserved.

One of the most robust bridges between the linear and the nonlinear, the simple and the complex, is scale analysis, the dimensional analysis of physical systems. It is through scale analysis that we can often make sense of complex nonlinear phenomena in terms of simpler models. At its core reside two questions. The first asks what quantities matter most to the problem at hand (which tends to be less obvious than one would like). The second asks what the expected magnitude and—importantly—dimensions of such quantities are. This second question is particularly important, as it captures the simple yet fundamental point that physical behavior should be invariant to the units we use to measure quantities. It may sound like an abstraction, but, without jargon, you could really call scale analysis "focusing systematically only on what matters most at a given time and place."

There are some subtle facts about scale analysis that make it more powerful than simply comparing orders of magnitude. A most remarkable example is that scale analysis can be applied, through a systematic use of dimensions, even when the precise equations governing the dynamics of a system are not known. The great physicist G. I. Taylor, a character whose prolific legacy haunts any aspiring scientist, gave a famous demonstration of this deceptively simple approach. In the 1950s, back when the detonating power of the nuclear bomb was a carefully guarded secret, the U.S. government incautiously released some unclassified photographs of a nuclear explosion. Taylor realized that whereas its details would be complex, the fundamentals of the problem would be governed by a few parameters. From dimensional arguments, he posited that there ought to be a scale-invariant number linking the radius of the blast, the time from detonation, energy released in the explosion, and the density of the surrounding air. From the photographs, he was able to estimate the radius and timing of the blast, inferring a remarkably accurate—and embarrassingly public—estimate of the energy of the explosion.

Taylor's capacity for insight was no doubt uncommon: Scale analysis seldom generates such elegant results. Nevertheless, it has a surprisingly wide range of applications and an illustrious history of guiding research in applied sciences, from structural engineering to turbulence theory.

But what of its broader application? The analysis of scales and dimensions can help us understand many complex problems and should be part of everybody's toolkit. In business planning and financial analysis, for example, the use of ratios and benchmarks is a first step toward scale analysis. It is certainly not a coincidence that they became common management tools at the height of Taylorism—a different Taylor, F. W. Taylor, the father of modern management theory—when "scientific management" and its derivatives made their first mark. The analogy is not without problems and would require further detailing than we have time for here—for example, on the use of dimensions to infer relations between quantities. But inventory turnover, profit margin, debt and equity ratios, labor and capital productivity, are dimensional parameters that could tell us a great deal about the basic dynamics of business economics, even without detailed market knowledge and day-to-day dynamics of individual transactions.

In fact, scale analysis in its simplest form can be applied to almost every quantitative aspect of daily life, from the fundamental time scales governing our expectations on returns on investments to the energy intensity of our lives. Ultimately, scale analysis is a particular form of numeracy—one where the relative magnitude as well as the dimensions of things that surround us guide our understanding of their meaning and evolution. It almost has the universality and coherence of Warburg's Mnemosyne Atlas: a unifying system of classification, where distant relations between seemingly disparate objects can continuously generate new ways of looking at problems and, through simile and dimension, can often reveal unexpected avenues of investigation.

Of course, any time a complicated system is translated into a simpler one, information is lost. Scale analysis is a tool only as insightful as the person using it. By itself, it does not provide answers and is no substitute for deeper analysis. But it offers a powerful lens through which to view reality and to understand "the order of things."

HIDDEN LAYERS

FRANK WILCZEK

Physicist, MIT; recipient, 2004 Nobel Prize in physics; author, The Lightness of Being: Mass, Ether, and the Unification of Forces

When I first took up the piano, merely hitting each note required my full attention. With practice, I began to work in phrases and chords. Eventually I made much better music with much less conscious effort.

Something powerful had happened in my brain.

That sort of experience is very common, of course. Something similar occurs whenever we learn a new language, master a new game, or get comfortable in a new environment. It seems very likely that a common mechanism is involved. I think it's possible to identify, in broad terms, what that mechanism is: We create *hidden layers.*

The scientific concept of a hidden layer arose from the study of neural networks. Here, a little picture is worth a thousand words:

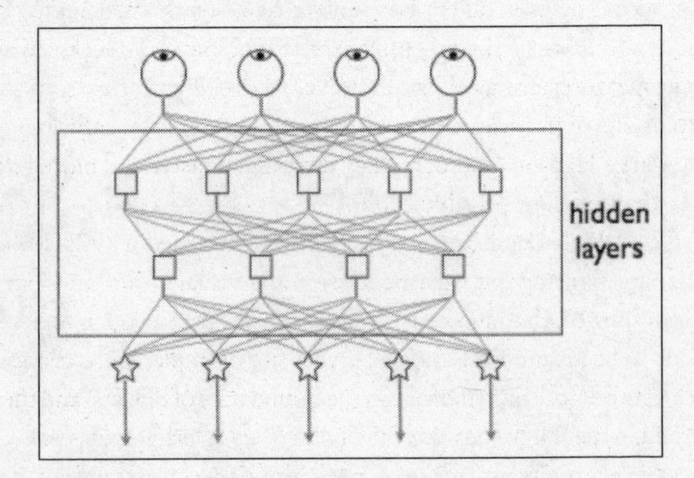

In this picture, the flow of information runs from top to bottom. Sensory neurons—the eyeballs at the top—take input from the external world and encode it into a convenient form (which is typically electrical pulse trains for biological neurons and numerical data for the computer "neurons" of artificial neural networks). They distribute this encoded information to other neurons, in the next layer below. Effector neurons—the stars at the bottom—send their signals to output devices (which are typically muscles for biological neurons and computer terminals for artificial neurons). In between are neurons that neither see nor act upon the outside world directly. These interneurons communicate only with other neurons. They are the hidden layers.

The earliest artificial neural networks lacked hidden layers. Their output was, therefore, a relatively simple function of their input. Those two-layer, input-output "perceptrons" had crippling limitations. For example, there is no way to design a perceptron that, faced with a series of different pictures of a few black circles on a white background, counts the number of circles. It took until the 1980s, decades after the pioneering work, for people to realize that including even one or two hidden layers could vastly enhance the capabilities of their neural networks. Nowadays such multilayer networks are used, for example, to distill patterns from the explosions of particles that emerge from high-energy collisions at the Large Hadron Collider. They do it much faster and more reliably than humans possibly could.

David Hubel and Torsten Wiesel were awarded a 1981 Nobel Prize for figuring out what neurons in the visual cortex are doing. They showed that successive hidden layers first extract features likely to be meaningful in visual scenes (for example, sharp changes in brightness or color, indicating the boundaries of objects) and then assemble them into meaningful wholes (the underlying objects).

In every moment of our adult waking life, we translate raw

patterns of photons hitting our retinas—photons arriving every which way from a jumble of unsorted sources and projected onto a two-dimensional surface—into the orderly, three-dimensional visual world we experience. Because it involves no conscious effort, we tend to take that everyday miracle for granted. But when engineers tried to duplicate it in robotic vision, they got a hard lesson in humility. Robotic vision remains today, by human standards, primitive. Hubel and Wiesel exhibited the architecture of nature's solution. It is the architecture of hidden layers.

Hidden layers embody in a concrete physical form the fashionable but rather vague and abstract idea of *emergence*. Each hidden-layer neuron has a template. The neuron becomes activated, and sends signals of its own to the next layer, precisely when the pattern of information it is receiving from the preceding layer matches (within some tolerance) that template. But this is just to say, in precision-enabling jargon, that the neuron defines, and thus creates, a new *emergent* concept.

In thinking about hidden layers, it's important to distinguish between the routine efficiency and power of a good network, once that network has been set up, and the difficult issue of how to set it up in the first place. That difference is reflected in the difference between playing the piano (or, say, riding a bicycle, or swimming) once you've learned (easy) and learning to do it in the first place (hard). Understanding exactly how new hidden layers get laid down in neural circuitry is a great unsolved problem of science. I'm tempted to say it's the greatest unsolved problem.

Liberated from its origin in neural networks, the concept of hidden layers becomes a versatile metaphor with genuine explanatory power. For example, in my work in physics I've noticed many times the impact of inventing names for things. When Murray Gell-Mann invented "quarks," he was giving a name to a paradoxical pattern of facts. Once that pattern was recognized, physicists faced the challenge of refining it into something mathematically precise and consistent, but identifying the problem was the crucial step toward solving it! Similarly, when I invented "anyons," for theoretical particles existing in only two dimensions, I knew I had put my finger on a coherent set of ideas, but I hardly anticipated how wonderfully those ideas would evolve and be embodied in reality. In cases like this, names create new nodes in hidden layers of thought.

I'm convinced that the general concept of hidden layers captures deep aspects of the way minds—whether human, animal, or alien; past, present, or future—do their work. Minds create useful concepts by embodying them in a specific way: namely, as features recognized by hidden layers. And isn't it pretty that "hidden layers" is itself a most useful concept, worthy to be included in hidden layers everywhere?

"SCIENCE"

LISA RANDALL

Physicist, Harvard University; author, Warped Passages: Unraveling the Mysteries of the Universe's Hidden Dimensions

The word "science" itself might be the best answer to this year's *Edge* Question. The idea that we can systematically understand certain aspects of the world and make predictions based on what we've learned, while appreciating and categorizing the extent and limitations of what we know, plays a big role in how we think. Many words that summarize the nature of science, such as "cause and effect," "predictions," and "experiments"—as well as words describing probabilistic results such as "mean," "median," "standard deviation," and the notion of "probability" itself—help us understand more specifically what we mean by "science" and how to interpret the world and the behavior within it.

"Effective theory" is one of the more important notions within science—and outside it. The idea is to determine what you can actually measure and decide, given the precision and accuracy of your measuring tools, and to find a theory appropriate to those measurable quantities. The theory that works might not be the ultimate truth, but it's as close an approximation to the truth as you need and is also the limit to what you can test at any given time. People can reasonably disagree on what lies beyond the effective theory, but in a domain where we have tested and confirmed it, we understand the theory to the degree that it's been tested.

An example is Newton's laws of motion, which work as well as we will ever need when they describe what happens to a ball when we throw it. Even though we now know that quantum mechanics is ultimately at play, it has no visible consequences on the trajectory of the ball. Newton's laws are part of an effective theory that is ultimately subsumed into quantum mechanics. Yet Newton's laws remain practical and true in their domain of validity. It's similar to the logic you apply when you look at a map. You decide the scale appropriate to your journey—are you traveling across the country, going upstate, or looking for the nearest grocery store?—and use the map scale appropriate to your question.

Terms that refer to specific scientific results can be efficient at times, but they can also be misleading when taken out of context and not supported by true scientific investigation. However, the scientific methods for seeking, testing, and identifying answers and understanding the limitations of what we have investigated will always be reliable ways of acquiring knowledge. A better understanding of the robustness and limitations of what science establishes, as well as those of probabilistic results and predictions, could make the world a place where people make the right decisions.

THE EXPANDING IN-GROUP

MARCEL KINSBOURNE
Neurologist and cognitive neuroscientist, The New School; coauthor (with Paula Kaplan), Children's Learning and Attention Problems

The ever-cumulating dispersion not only of information but also of population across the globe is the great social phenomenon of this age. Regrettably, cultures are being homogenized, but cultural differences are also being demystified, and intermarriage is escalating, across ethnic groups within states and among ethnicities across the world. The effects are potentially beneficial for the improvement of cognitive skills, from two perspectives. We can call these the "expanding in-group" and the "hybrid vigor" effects.

The in-group-vs.-out-group double standard, which had and has such catastrophic consequences, could in theory be eliminated if everyone alive were considered to be in everyone else's in-group. This utopian prospect is remote, but an expansion of the conceptual in-group would expand the range of friendly, supportive, and altruistic behavior. This effect may already be in evidence in the increase in charitable activities in support of foreign populations confronted by natural disasters. Donors identifying to a greater extent with recipients make this possible. The rise in international adoptions also indicates that the barriers set up by discriminatory and nationalistic prejudice are becoming porous.

The other potential benefit is genetic. The phenomenon of hybrid vigor in offspring, which is also called heterozygote advantage, derives from a cross between dissimilar parents. It is well established experimentally, and the benefits of mingling disparate gene pools are seen not only in improved physical but also improved mental development. Intermarriage therefore promises cognitive benefits. Indeed, it may already have contributed to the Flynn effect, the well-known worldwide rise in average measured intelligence by as much as three IQ points per decade over successive decades since the early twentieth century.

Every major change is liable to unintended consequences. These can be beneficial, detrimental, or both. The social and cognitive benefits of the intermingling of people and populations are no exception, and there is no knowing whether the benefits are counterweighed or even outweighed by as yet unknown drawbacks. Nonetheless, unintended though they might be, the social benefits of the overall greater probability of in-group status, and the cognitive benefits of increasing frequency of intermarriage entailed by globalization, may already be making themselves felt.

CONTINGENT SUPERORGANISMS

JONATHAN HAIDT
Social psychologist, University of Virginia; author, The Happiness Hypothesis

Humans are the giraffes of altruism. We're freaks of nature, able (at our best) to achieve antlike levels of service to the group. We readily join together to create superorganisms, but unlike the eusocial insects we do it with blatant disregard for kinship and we do it temporarily and contingent upon special circumstances (particularly intergroup conflict, as is found in war, sports, and business).

Ever since the publication of G. C. Williams's 1966 classic *Adaptation and Natural Selection*, biologists have joined with social scientists to form an altruism debunkery society. Any human or animal act that appears altruistic has been explained away as selfishness in disguise, linked ultimately to kin selection (genes help copies of themselves) or reciprocal altruism (agents help only to the extent that they can expect a positive return, including to their reputations).

But in the last few years there's been a growing acceptance of the idea that "Life is a self-replicating hierarchy of levels" and natural selection operates on multiple levels simultaneously, as Bert Hölldobler and E. O. Wilson put it in their recent book, *The Superorganism*. Whenever the free-rider problem is solved at one level of the hierarchy, such that individual agents can link their fortunes and live or die as a group, a superorganism is formed. Such "major transitions" are rare in the history of life, but when they have happened, the resulting superorganisms have been wildly successful. (Eukaryotic cells, multicelled organisms, and ant colonies are all examples of such transitions).

Building on Hölldobler and Wilson's work on insect societies, we can define a "contingent superorganism" as a group of people that forms a functional unit in which each member is willing to sacrifice for the good of the group in order to surmount a challenge or threat, usually from another contingent superorganism. It is the most noble and the most terrifying human ability. It is the secret of successful hivelike organizations, from the hierarchical corporations of the 1950s to the more fluid dot-coms of today. It is the purpose of basic training in the military. It is the reward that makes people want to join fraternities, fire departments, and rock bands. It is the dream of fascism.

Having the term "contingent superorganism" in our cognitive toolkit may help people to overcome forty years of biological reductionism and gain a more accurate view of human nature, human altruism, and human potential. It can explain our otherwise freakish love of melding ourselves (temporarily, contingently) into something larger than ourselves.

THE PARETO PRINCIPLE

CLAY SHIRKY
Social and technology network topology researcher; adjunct professor, NYU Graduate School of Interactive Telecommunications Program; author, Cognitive Surplus: Creativity and Generosity in a Connected Age

You see the pattern everywhere: The top 1 percent of the population controls 35 percent of the wealth. On Twitter, the top 2 percent of users sends 60 percent of the messages. In the health-care system, the treatment of the most expensive fifth of patients creates four-fifths of the overall cost. These figures are always reported as shocking, as if the normal order of things had been disrupted, as if the appearance of anything other than a completely linear distribution of money or messages or effort were a surprise of the highest order.

It's not. Or rather, it shouldn't be.

The Italian economist Vilfredo Pareto undertook a study of market economies a century ago and discovered that no matter what the country, the richest quintile of the population controlled most of the wealth. The effects of this Pareto distribution go by many names—the *80/20 rule*, *Zipf's law*, the *power-law* distribution, *winner-take-all*—but the basic shape of the underlying distribution is always the same: The richest or busiest or most connected participants in a system will account for much, *much* more wealth or activity or connectedness than average.

Furthermore, this pattern is recursive. Within the top 20 percent of a system that exhibits a Pareto distribution, the top 20 percent of *that* slice will also account for disproportionately more of whatever is being measured, and so on. The most highly ranked element of such a system will be much more highly weighted than even the #2 item in the same chart. (The word "the" is not only the commonest word in English, it appears twice as often as the second most common, "of.")

This pattern was so common that Pareto called it a "predictable imbalance"; despite this bit of century-old optimism, however, we are still failing to predict it, even though it is everywhere.

Part of our failure to expect the expected is that we have been taught that the paradigmatic distribution of large systems is the Gaussian distribution, commonly known as the bell curve. In a bell-curve distribution—like height, say—the average and the median (the middle point in the system) are the same. The average height of a hundred American women selected at random will be about 5'4" and the height of the fiftieth-ranked woman will also be 5'4".

Pareto distributions are nothing like that: The recursive 80/20 weighting means that the average is far from the middle. This in turn means that in such systems most people (or whatever is being measured) are below average, a pattern encapsulated in the old economics joke: "Bill Gates walks into a bar and makes everybody a millionaire, on average."

The Pareto distribution shows up in a remarkably wide array of complex systems. Together, "the" and "of" account for 10 percent of all words used in English. The most volatile day in the history of a stock market will typically be twice as volatile as that of the second-most volatile and ten times the tenth-most. Tag frequency on Flickr photos obeys a Pareto distribution, as does the magnitude of earthquakes, the popularity of books, the size of asteroids, and the social connectedness of your friends. The Pareto principle is so basic to the sciences that special graph paper showing Pareto distributions as straight lines rather than as steep curves is manufactured by the ream.

And yet, despite a century of scientific familiarity, samples drawn from Pareto distributions are routinely presented to the public as anomalies, which prevents us from thinking clearly about the world. We should stop thinking that average family income and the income of the median family have anything to do with one another, or that enthusiastic and normal users of communications tools are doing similar things, or that extroverts should be only moderately more connected than normal people. We should stop thinking that the largest future earthquake or market panic will be as large as the largest historical one; the longer a system persists, the likelier it is that an event twice as large as all previous ones is coming.

This doesn't mean that such distributions are beyond our ability to affect them. A Pareto curve's decline from head to tail can be more or less dramatic, and in some cases, political or social intervention can affect that slope—tax policy can raise or lower the share of income of the top 1 percent of a population, just as there are ways to constrain the overall volatility of markets, or reduce the band in which health-care costs can fluctuate.

However, until we assume such systems *are* Pareto distributions and will remain so even after any such intervention, we haven't even started thinking about them in the right way. In all likelihood, we're trying to put a Pareto peg in a Gaussian hole. A hundred years after the discovery of this predictable imbalance, we should finish the job and actually start expecting it.

FIND THAT FRAME

WILLIAM CALVIN

Theoretical neurobiologist; emeritus professor, University of Washington School of Medicine; author, Global Fever: How to Treat Climate Change

An automatic stage of "compare and contrast" would improve most cognitive functions, not just the grade on an essay. You set up a comparison—say, that the interwoven melodies of rock and roll are like how you must twist when dancing on a boat when the bow is rocking up and down in a different rhythm than the deck is rolling from side to side.

Comparison is an important part of trying ideas on for size, for finding related memories and exercising constructive skepticism. Without it, you can become trapped in someone else's framing of a problem. You often need to know where someone is coming from—and while Compare 'n' Contrast is your best friend, you may also need to search for the cognitive framing. What has been cropped out of the frame can lead the unwary to an incorrect inference, as when they assume that what is left out is unimportant. For example, "We should reach a 2°C (3.6°F) fever in the year 2049" always makes me want to interject "Unless another abrupt climate shift gets us there next year."

Global warming's ramp-up in temperature is the aspect of climate change that climate scientists can currently calculate—that's where they are coming from. And while this can produce really important insights—even big emission reductions only delay the 2°C fever for nineteen years—it leaves out all of those abrupt climate shifts observed since 1976, as when the world's drought acreage doubled in 1982 and jumped from double to triple in 1997, then back to double in 2005. That's like stairs, not a ramp.

Even if we thoroughly understood the mechanism for an abrupt climate shift—likely a rearrangement of the winds that produce Deluge 'n' Drought by delivering ocean moisture elsewhere, though burning down the Amazon rain forest should also trigger a big one—chaos theory's butterfly effect says we still could not predict when a big shift will occur or what size it will be. That makes a climate surprise like a heart attack. You can't predict when. You can't say whether it will be minor or catastrophic. But you can often prevent it—in the case of climate, by cleaning up the excess CO_2.

Drawing down the CO_2 is also typically excluded from the current climate framing. Mere emissions reduction now resembles locking the barn door after the horse is gone—worthwhile, but not exactly recovery. Politicians usually love locking barn doors, as it gives the appearance of taking action cheaply. Emissions reduction only slows the rate at which things get worse, as the CO_2 accumulation keeps growing. (People confuse annual emissions with the accumulation that causes the trouble.) On the other hand, cleaning up the CO_2 actually cools things, reverses ocean acidification, and even reverses the thermal-expansion portion of rising sea level.

Recently I heard a biologist complaining about models for insect social behavior: "All of the difficult stuff is not mentioned. Only the easy stuff is calculated." Scientists first do what they already know how to do. But their quantitative results are no substitute for a full qualitative account. When something is left out because it is computationally intractable (sudden shifts) or would just be a guess (cleanup), they often don't bother to mention it at all. "Everybody [in our field] knows that" just won't do when people outside the field are hanging on your every word.

So find that frame and ask about what was left out. Like abrupt climate shifts or a CO_2 cleanup, it may be the most important consideration of all.

WICKED PROBLEMS

JAY ROSEN
Professor of journalism, New York University; author, What Are Journalists For?

There's a problem that anyone who has lived in New York City must wonder about: You can't get a cab from 4:00 to 5:00 P.M. The reason for this is not a mystery: At a moment of peak demand, taxi drivers tend to change shifts. Too many cabs are headed to garages in Queens, because when a taxi is operated by two drivers twenty-four hours a day, a fair division of shifts is to switch over at 5:00 P.M. Now, this is a problem for the city's Taxi and Limousine Commission, it may even be a hard one to solve, but it is not a wicked problem. For one thing, it's easy to describe, as I just showed you. That, right there, boots it from the wicked category.

Among some social scientists, there is this term of art: *wicked problems.* We would be vastly better off if we understood what wicked problems are and learned to distinguish between them and regular (or "tame") problems.

Wicked problems have these features: It is hard to say what the problem is, to define it clearly, or to tell where it stops and starts. There is no "right" way to view the problem, no definitive formulation. The way it's framed will change what the solution appears to be. Someone can always say that the problem is just a symptom of another problem, and that someone will not be wrong. There are many stakeholders, all with their own frames, which they tend to see as exclusively correct. Ask what the problem is and you will get a different answer from each. The problem is interconnected to a lot of other problems; pulling them apart is almost impossible.

It gets worse. Every wicked problem is unique, so in a sense there is no prior art, and solving one won't help you with the others. No one has "the right to be wrong"; meaning has enough legitimacy and stakeholder support to try stuff that will almost certainly fail at first. Instead, failure is savaged, and the trier is deemed unsuitable for another try. The problem keeps changing on us. It is never definitely resolved. We just run out of patience, or time, or money. It's not possible to understand the problem first, then solve it; rather, attempts to solve it reveal further dimensions of the problem. (Which is the secret of success for people who are "good" at wicked problems.)

Know any problems like that? Sure you do. Probably the best example in our time is climate change. What could be more interconnected than it? Someone can always say that climate change is just a symptom of another problem—our entire way of life, perhaps—and he or she would not be wrong. We've certainly never solved anything like it before. Stakeholders: everyone on the planet, every nation, every company.

When General Motors was about to go bankrupt and throw tens of thousands of people out of work, that was a big, honking problem, which rightly landed on the president's desk, but it was not a wicked one. Barack Obama's advisors could present him with a limited range of options; if he decided to take the political risk and save General Motors from collapse, he could be reasonably certain that the recommended actions would work. If they didn't, he could try more drastic measures.

But health care reform wasn't like that at all. In the United States, rising health care costs are a classic case of a wicked problem. No "right" way to view it. Every solution comes with its own contestable frame. There are multiple stakeholders, who don't define the problem the same way. If the number of uninsured goes down but costs go up, is that progress? We don't even know.

Wicked!

Still, we would be better off if we knew when we were dealing with a wicked problem as opposed to the regular kind. If we could designate some problems as wicked, we might realize that "normal" approaches to problem solving don't work. We can't define the problem, evaluate possible solutions, pick the best one, hire the experts, and implement. No matter how much we may want to follow a routine like that, it won't succeed. Institutions may require it, habit may favor it, the boss may order it, but wicked problems don't care.

Presidential debates that divided wicked from tame problems would be very different debates. Better, I think. Journalists who covered wicked problems differently from the way in which they covered normal problems would be smarter journalists. Institutions that knew how to distinguish wicked problems from the other kind would eventually learn the limits of command and control.

Wicked problems demand people who are creative, pragmatic, flexible, and collaborative. They never invest too much in their ideas, because they know they will have to alter them. They know there's no right place to start, so they simply start somewhere and see what happens. They accept the fact that they're more likely to understand the problem after it's solved than before. They don't expect to get a good solution; they keep working until they've found something that's good enough. They're never convinced they know enough to solve the problem, so they're constantly testing their ideas on different stakeholders.

Know any people like that? Maybe we can get them interested in health care . . .

ANTHROPOCENE THINKING

DANIEL GOLEMAN
Psychologist; author, Emotional Intelligence

Do you know the PDF of your shampoo? A PDF refers to a "partially diminished fraction" of an ecosystem, and if your shampoo contains palm oil cultivated on clear-cut jungle in Borneo, say, that value will be high. How about your shampoo's DALY? This measure comes from public health: "disability-adjusted life years," or the amount of one's life that will be lost to a disabling disease because of, say, a lifetime's cumulative exposure to a given industrial chemical. So if your favorite shampoo contains two common ingredients, the carcinogen 1,4-Dioxane, or BHA, an endocrine disrupter, its DALY will be higher.

PDFs and DALYs are among myriad metrics for anthropocene thinking, which views how human systems affect the global systems that sustain life. This way of perceiving interactions between the built and the natural worlds comes from the geological sciences. If adopted more widely, this lens might usefully inform how we find solutions to the singular peril our species faces: the extinction of our ecological niche.

Beginning with cultivation and accelerating with the Industrial Revolution, our planet left the Holocene epoch and entered what geologists call the Anthropocene, in which human systems erode the natural systems that support life. Through the Anthropocene lens, the daily workings of the energy grid, transportation, industry, and commerce inexorably deteriorate global biogeochemical systems, such as the carbon, phosphorus, and water cycles. The most troubling data suggest that since the 1950s the human enterprise has led to an explosive acceleration that will reach criticality within the next few decades, as different systems reach a tipping point-of-no-return. For instance, about half the total rise in atmospheric CO_2 concentration has occurred in just the last thirty years—and of all the global life-support systems, the carbon cycle is closest to no-return. While such "inconvenient truths" about the carbon cycle have been the poster child for our species' slow-motion suicide, that's just part of a much larger picture, with all the eight global life-support systems under attack by our daily habits.

Anthropocene thinking tells us the problem is not necessarily inherent in the systems like commerce and energy that degrade nature; one hopes that these can be modified to become self-sustaining, with innovative advances and entrepreneurial energy. The real root of the Anthropocene dilemma lies in our neural architecture.

We approach the Anthropocene threat with brains shaped by evolution to survive the previous geological epoch, the Holocene, when dangers were signaled by growls and rustles in the bushes, and it served one well to reflexively abhor spiders and snakes. Our neural alarm systems still attune to this largely antiquated range of danger.

Add to that misattunement to threats our built-in perceptual blind spot: We have no direct neural register for the dangers of the Anthropocene epoch, which are too macro or micro for our sensory apparatus. We are oblivious to, say, our body burden, the lifetime build-up of damaging industrial chemicals in our tissues.

To be sure, we have methods for assessing CO_2 buildups or blood levels of BHA. But for the vast majority of people, those numbers have little to no emotional impact. Our amygdala shrugs.

Finding ways to counter the forces that feed the Anthropocene effect should count high in prioritizing scientific efforts. The earth sciences, of course, embrace the issue, but they do not deal with the root of the problem—human behavior. The sciences that have most to offer have done the least Anthropocene thinking.

The fields that hold keys to solutions include economics, neuroscience, social psychology, and cognitive science, and their various hybrids. With a focus on Anthropocene theory and practice, they might well contribute species-saving insights. But first they have to engage this challenge, which for the most part has remained off their agenda.

When, for example, will neuroeconomics tackle the brain's perplexing indifference to the news about planetary meltdown, let alone how that neural blind spot might be patched? Might cognitive neuroscience one day offer some insight that would change our collective decision making away from a lemming's march to oblivion? Could any of the computer, behavioral, or brain sciences come up with an information prosthetic that might reverse our course?

Paul Crutzen, the Dutch atmospheric chemist who received a Nobel for his work on ozone depletion, coined the term "Anthropocene" ten years ago. As a meme, "Anthropocene" has as yet little traction in scientific circles beyond geology and environmental science, let alone the wider culture: A Google check on "anthropocene," as of this writing, shows 78,700 references (mainly in geoscience), while, by contrast, "placebo," a once esoteric medical term now well established as a meme, has more than 18 million (and even the freshly coined "vuvuzela" has 3,650,000).

HOMO DILATUS

ALUN ANDERSON

Senior consultant, former editor-in-chief and publishing director, New Scientist; *author,* After the Ice: Life, Death, and Geopolitics in the New Arctic

Our species might well be renamed *Homo dilatus*, the procrastinating ape. Somewhere in our evolution, we acquired the brain circuitry to deal with sudden crises and respond with urgent action. Steady declines and slowly developing threats are quite different. "Why act now, when the future is far off?" is the maxim for a species designed to deal with near-term problems and not long-term uncertainties. It's a handy view of humankind that all those who use science to change policy should keep in their mental toolkit, and a tendency greatly reinforced by the endless procrastination in tackling climate change. Cancún follows Copenhagen follows Kyoto, but the more we dither and no extraordinary disaster follows, the more dithering seems just fine.

Such behavior is not unique to climate change. It took the sinking of the *Titanic* to put sufficient lifeboats on passenger ships, the huge spill from the *Amoco Cadiz* to set international marine pollution rules, and the *Exxon Valdez* disaster to drive the switch to double-hulled tankers. The same pattern is seen in the oil industry, with the 2010 Gulf spill the latest chapter in the "Disaster first; regulations later" mind-set of *Homo dilatus*.

There are a million similar stories from human history. So many great powers and once-dominant corporations slipped away as their fortunes declined without the necessary crisis to force change. Slow and steady change simply leads to habituation, not action. You could walk in the British countryside now and hear only a fraction of the birdsong that would have delighted a Victorian poet, but we simply cannot feel insidious loss. Only a present crisis wakes us.

So puzzling is our behavior that the "psychology of climate change" has become a significant area of research, with efforts to find those vital messages that will turn our thinking toward the longer term and away from the concrete Now. Sadly, the skull of *Homo dilatus* seems too thick for the tricks that are currently on offer. In the case of climate change, we might better focus on adaptation until a big crisis comes along to rivet our minds. The complete loss of the summer Arctic ice might be the first. A huge dome of shining ice, about half the size of the United States, covers the top of the world in summer now. In a couple of decades, it will likely be gone. Will millions of square kilometers of white ice turning to dark water feel like a crisis? If that doesn't do it, then following soon after will likely be painful and persistent droughts across the United States, much of Africa, Southeast Asia, and Australia.

Then the good side of *Homo dilatus* may finally surface. A crisis might bring out the Bruce Willis in all of us, and with luck we'll find an unexpected way to right the world before the end of the reel. Then we'll no doubt put our feet up again.

WE ARE LOST IN THOUGHT

SAM HARRIS

Neuroscientist; chairman, Project Reason; author, The Moral Landscape *and* The End of Faith

I invite you to pay attention to anything—the sight of this text, the sensation of breathing, the feeling of your body resting against your chair—for a mere sixty seconds without getting distracted by discursive thought. It sounds simple enough: Just pay attention. The truth, however, is that you will find the task impossible. If the lives of your children depended on it, you could not focus on anything—even the feeling of a knife at your throat—for more than a few seconds, before your awareness would be submerged again by the flow of thought. This forced plunge into unreality is a problem. In fact, it is the problem from which every other problem in human life appears to be made.

I am by no means denying the importance of thinking. Linguistic thought is indispensable to us. It is the basis for planning, explicit learning, moral reasoning, and many other capacities that make us human. Thinking is the substance of every social relationship and cultural institution we have. It is also the foundation of science. But our habitual identification with the flow of thought—that is, our failure to recognize thoughts *as thoughts*, as transient appearances in consciousness—is a primary source of human suffering and confusion.

Our relationship to our own thinking is strange to the point of paradox, in fact. When we see a person walking down the street talking to himself, we generally assume that he is mentally ill. But we all talk to ourselves *continuously*—we just have the good sense to keep our mouths shut. Our lives in the present can scarcely be glimpsed through the veil of our discursivity: We tell ourselves what just happened, what almost happened, what should have happened, and what might yet happen. We ceaselessly reiterate our hopes and fears about the future. Rather than simply existing as ourselves, we seem to presume a relationship with ourselves. It's as though we were having a conversation with an imaginary friend possessed of infinite patience. Who are we talking to?

While most of us go through life feeling that we are the thinker of our thoughts and the experiencer of our experience, from the perspective of science we know that this is a distorted view. There is no discrete self or ego lurking like a Minotaur in the labyrinth of the brain. There is no region of cortex or pathway of neural processing that occupies a privileged position with respect to our personhood. There is no unchanging "center of narrative gravity" (to use Daniel Dennett's phrase). In subjective terms, however, there *seems* to be one—to most of us, most of the time.

Our contemplative traditions (Hindu, Buddhist, Christian, Muslim, Jewish, etc.) also suggest, to varying degrees and with greater or lesser precision, that we live in the grip of a cognitive illusion. But the alternative to our captivity is almost always viewed through the lens of religious dogma. A Christian will recite the Lord's Prayer continuously over a weekend, experience a profound sense of clarity and peace, and judge this mental state to be fully corroborative of the doctrine of Christianity; a Hindu will spend an evening singing devotional songs to Krishna, feel suddenly free of his conventional sense of self, and conclude that his chosen deity has showered him with grace; a Sufi will spend hours whirling in circles, pierce the veil of thought for a time, and believe that he has established a direct connection to Allah.

The universality of these phenomena refutes the sectarian claims of any one religion. And, given that contemplatives generally present their experiences of self-transcendence as inseparable from their associated theology, mythology, and metaphysics, it is no surprise that scientists and nonbelievers tend to view their reports as the product of disordered minds, or as exaggerated accounts of far more common mental states—like scientific awe, aesthetic enjoyment, artistic inspiration, and so on.

Our religions are clearly false, even if certain classically religious experiences are worth having. If we want to actually understand the mind, and overcome some of the most dangerous and enduring sources of conflict in our world, we must begin thinking about the full spectrum of human experience in the context of science.

But we must first realize that we are lost in thought.

THE PHENOMENALLY TRANSPARENT SELF-MODEL

THOMAS METZINGER
Philosopher, Johannes Gutenberg-Universität, Mainz, and Frankfurt Institute for Advanced Studies; author, The Ego Tunnel

A self-model is the inner representation that some information-processing systems have of themselves as a whole. A representation is phenomenally transparent if it (a) is conscious and (b) cannot be experienced *as* a representation. Therefore, transparent representations create the phenomenology of naïve realism—the robust and irrevocable sense that you are directly and immediately perceiving something that must be real. Now apply the second concept to the first: A transparent self-model necessarily creates the realistic conscious experience of selfhood—of being directly and immediately in touch with oneself as a whole.

This concept is important, because it shows how, in a certain class of information-processing systems, the robust phenomenology of *being a self* would inevitably appear—although these systems never were, or had, anything like a self. It is empirically plausible that we might just be such systems.

CORRELATION IS NOT A CAUSE

SUE BLACKMORE

Psychologist; author, Consciousness: An Introduction

The sentence "Correlation is not a cause" (CINAC) may be familiar to scientists but has not found its way into everyday language, even though critical thinking and scientific understanding would improve if more people had this simple reminder in their mental toolkit.

One reason for this lack is that CINAC can be surprisingly difficult to grasp. I learned just how difficult when I was teaching experimental design to nurses, physiotherapists, and other assorted groups. They usually understood my favorite example: Imagine you're watching at a railway station. More and more people arrive, until the platform is crowded, and then—hey, presto!—along comes a train. Did the people cause the train to arrive (*A* causes *B*)? Did the train cause the people to arrive (*B* causes *A*)? No, they both depended on a railway timetable (*C* caused both *A* and *B*).

I soon discovered that this understanding tended to slip away again and again, until I began a new regimen and started every lecture with an invented example to get them thinking. "Right," I might say. "Suppose it's been discovered—I don't mean it's true—that children who eat more tomato ketchup do worse in their exams. Why could this be?"

They would argue that it wasn't true. (I'd explain the point of thought experiments again.)

"But there'd be health warnings on ketchup if it were poisonous."

(Just pretend it's true for now, please.)

And then they'd start using their imaginations: "There's something in the ketchup that slows down nerves." "Eating ketchup makes you watch more telly instead of doing your homework." "Eating more ketchup means eating more chips, and that makes you fat and lazy."

Yes, yes, probably wrong but great examples of *A* causes *B*. Go on.

And so to "Stupid people have different taste buds and don't like ketchup." "Maybe if you don't pass your exams, your mum gives you ketchup." And finally "Poorer people eat more junk food and do less well at school."

Next week: "Suppose we find that the more often people consult astrologers or psychics, the longer they live."

"But it can't be true—astrology's bunkum."

(Sigh . . . just pretend it's true for now, please.)

OK. "Astrologers have a special psychic energy that they radiate to their clients." "Knowing the future means you can avoid dying." "Understanding your horoscope makes you happier and healthier."

Yes, yes, excellent ideas, go on.

"The older people get, the more often they go to psychics." "Being healthy makes you more spiritual, and so you seek out spiritual guidance."

Yes, yes, keep going, all testable ideas.

And finally "Women go to psychics more often and also live longer than men."

The point is that once you greet any new correlation with CINAC, your imagination is let loose. Once you listen to every new science story CINACally (which conveniently sounds like "cynically"), you find yourself thinking "OK, if *A* doesn't cause *B*, could *B* cause *A*? Could something else cause them both, or could they both be the same thing even though they don't appear to be? What's going on? Can I imagine other possibilities? Could I test them? Could I find out which is true?" Then you can be critical of the science stories you hear. Then you are thinking like a scientist.

Stories of health scares and psychic claims may get people's attention, but understanding that a correlation is not a cause could raise levels of debate over some of today's most pressing scientific issues. For example, we know that global temperature rise correlates with increasing levels of atmospheric carbon dioxide, but why? Thinking CINACally means asking which variable causes which, or whether something else causes both, with important consequences for social action and the future of life on Earth.

Some say that the greatest mystery facing science is the nature of consciousness. We *seem* to be independent selves having consciousness and free will, and yet the more we understand how the brain works, the less room there seems to be for consciousness to do anything. A popular way of trying to solve the mystery is the hunt for the "neural correlates of consciousness." For example, we know that brain activity in parts of the motor cortex and frontal lobes correlates with conscious decisions to act. But do our conscious decisions cause the brain activity, does the brain activity cause our decisions, or are both caused by something else?

The fourth possibility is that brain activity and conscious experiences are really the same thing, just as light turned out not to be *caused* by electromagnetic radiation but to *be* electromagnetic radiation, and heat turned out to be the movement of molecules in a fluid. At the moment, we have no inkling of how consciousness could *be* brain activity, but my guess is that it will turn out that way. Once we clear away some of our delusions about the nature of our own minds, we may finally see why there is no deep mystery and our conscious experiences simply *are* what is going on inside our brains. If this is right, then there are no neural correlates of consciousness. But whether it's right or not, remembering CINAC and working slowly from correlations to causes is likely to be how this mystery is finally solved.

INFORMATION FLOW

DAVID DALRYMPLE
Researcher, MIT Media Lab

The concept of cause-and-effect is better understood as the flow of information between two connected events, from the earlier event to the later one. Saying "*A* causes *B*" sounds precise but is actually very vague. I would specify much more by saying, "With the information that *A* has happened, I can compute with almost total confidence* that *B* will happen." This rules out the possibility that other factors could prevent *B* even if *A* does happen, but allows the possibility that other factors could cause *B* even if *A* doesn't happen.

As shorthand, we can say that one set of information "specifies" another, if the latter can be deduced or computed from the former. Note that this doesn't apply only to one-bit sets of information, like the occurrence of a specific event. It can also apply to symbolic variables (given the state of the Web, the results you get from a search engine are specified by your query), numeric variables (the number read off a precise thermometer is specified by the temperature of the sensor), or even behavioral variables (the behavior of a computer is specified by the bits loaded in its memory).

But let's take a closer look at the assumptions we're making. Astute readers may have noticed that in one of my examples, I

---

* In our universe, too many things are interconnected for absolute statements of any kind, so we usually relax our criteria; for instance, "total confidence" might be relaxed from a 0 percent chance of being wrong to, say, a 1 in 3 quadrillion chance of being wrong—about the chance that as you finish this sentence, all of humanity will be wiped out by a meteor.

assumed that the entire state of the Web was a constant. How ridiculous! In mathematical parlance, assumptions are known as "priors," and in a certain widespread school of statistical thought they are considered the most important aspect of any process involving information. What we really want to know is if, given a set of existing priors, adding one piece of information (*A*) would allow us to update our estimate of the likelihood of another piece of information (*B*). Of course, this depends on the priors—for instance, if our priors include absolute knowledge of *B*, then an update will not be possible.

If, for most reasonable sets of priors, information about *A* would allow us to update our estimate of *B*, then it would seem there is some sort of causal connection between the two. But the form of the causal connection is unspecified—a principle often stated as "correlation does not imply causation." The reason for this is that the essence of causation as a concept rests on our tendency to have information about earlier events before we have information about later events. (The full implications of this concept for human consciousness, the second law of thermodynamics, and the nature of time are interesting, but sadly outside the scope of this essay.)

If information about all events always came in the order in which the events occurred, then correlation would indeed imply causation. But in the real world, not only are we limited to observing events in the past but also we may discover information about those events out of order. Thus, the correlations we observe could be reverse causes (information about *A* allows us to update our estimate of *B*, although *B* happened first and thus was the cause of *A*) or even more complex situations (e.g., information about *A* allows us to update our estimate of *B* but is also giving us information about *C*, which happened before either *A* or *B* and caused both).

Information flow is symmetric: If information about *A* were to allow us to update our estimate of *B*, then information about *B* would allow us to update our estimate of *A*. But since we cannot change the past or know the future, these constraints are useful to us only when contextualized temporally and arranged in order of occurrence. Information flow is always from the past to the future, but in our minds some of the arrows may be reversed. Resolving this ambiguity is essentially the problem that science was designed to solve. If you can master the technique of visualizing all information flow and keeping track of your priors, then the full power of the scientific method—and more—is yours to wield from your personal cognitive toolkit.

THINKING IN TIME VERSUS THINKING OUTSIDE OF TIME

LEE SMOLIN

Physicist, Perimeter Institute; author, The Trouble with Physics

One very old and pervasive habit of thought is to imagine that the true answer to whatever question we are wondering about lies out there in some eternal domain of "timeless truths." The aim of research is then to "discover" the answer or solution in that already existing timeless domain. For example, physicists often speak as if the final theory of everything already exists in a vast timeless Platonic space of mathematical objects. This is thinking outside of time.

Scientists are thinking in time when we conceive of our task as the invention of genuinely novel ideas to describe newly discovered phenomena and novel mathematical structures to express them. If we think outside of time, we believe these ideas somehow "existed" before we invented them. If we think in time, we see no reason to presume that.

The contrast between thinking in time and thinking outside of time can be seen in many domains of human thought and action. We are thinking outside of time when, faced with a technological or social problem to solve, we assume the possible approaches are already determined by a set of absolute preexisting categories. We are thinking in time when we understand that progress in technology, society, and science happens by the invention of genuinely novel ideas, strategies, and novel forms of social organization.

The idea that truth is timeless and resides outside the universe was the essence of Plato's philosophy, exemplified in the parable of the slave boy, which was meant to argue that discovery is merely remembering. This is reflected in the philosophy of mathematics called Platonism, which is the belief that there are two ways of existing: Regular physical things exist in the universe and are subject to time and change, whereas mathematical objects exist in a timeless realm. The division of the world into a time-drenched earthly realm of life, death, change, and decay, surrounded by a heavenly sphere of perfect eternal truth, framed both ancient science and Christian religion.

If we imagine that the task of physics is the discovery of a timeless mathematical object that is isomorphic to the history of the world, then we imagine that the truth to the universe lies outside the universe. This is such a familiar habit of thought that we fail to see its absurdity: If the universe is all that exists, then how can something exist outside of it for it to be isomorphic to?

On the other hand, if we take the reality of time as evident, then there can be no mathematical object that is perfectly isomorphic to the world, because one property of the real world that is not shared by any mathematical object is that it is always some moment. Indeed, as Charles Sanders Peirce first observed, the hypothesis that the laws of physics evolved through the history of the world is necessary if we are to have a rational understanding of why one particular set of laws holds, rather than another.

Thinking outside of time often implies the existence of an imagined realm, outside the universe, where the truth lies. This is a religious idea, because it means that explanations and justifications ultimately refer to something outside the world we experience ourselves to be a part of. If we insist that there is nothing outside the universe, not even abstract ideas or mathematical objects, we are forced to find the causes of phenomena entirely within our universe. So thinking in time is also thinking within the one universe of phenomena our observations show us to inhabit.

Among contemporary cosmologists and physicists, proponents of eternal inflation and timeless quantum cosmology are thinking outside of time. Proponents of evolutionary and cyclic cosmological scenarios are thinking in time. If you think in time, you worry about time ending at space-time singularities. If you think outside of time, this is an ignorable problem, because you believe reality is the whole history of the world at once.

Darwinian evolutionary biology is the prototype for thinking in time, because at its heart is the realization that natural processes developing in time can lead to the creation of genuinely novel structures. Even novel laws can emerge when the structures to which they apply come to exist. Evolutionary dynamics has no need of abstract and vast spaces like all the possible viable animals, DNA sequences, sets of proteins, or biological laws. Exaptations are too unpredictable and too dependent on the whole suite of living creatures to be analyzed and coded into properties of DNA sequences. Better, as the theoretical biologist Stuart Kauffman proposes, to think of evolutionary dynamics as the exploration, in time, by the biosphere, of the *adjacent possible*.

The same goes for the evolution of technologies, economies, and societies. The poverty of the conception that economic markets tend to unique equilibria, independent of their histories, shows the danger of thinking outside of time. Meanwhile the path dependence that economist Brian Arthur and others show is necessary to understand real markets illustrates the kind of insights that are gotten by thinking in time.

Thinking in time is not relativism; it is a form of relationalism. Truth can be both time-bound and objective, when it is about objects that only exist once they are invented, by evolution or human thought.

When we think in time, we recognize the human ability to invent genuinely novel constructions and solutions to problems.

When we think about the organizations and societies we live and work in outside of time, we unquestioningly accept their strictures and seek to manipulate the levers of bureaucracy as if they had an absolute reason to be there. When we think about organizations in time, we recognize that every feature of them is a result of their history and everything about them is negotiable and subject to improvement by the invention of novel ways of doing things.

NEGATIVE CAPABILITY IS A PROFOUND THERAPY

RICHARD FOREMAN
Playwright and director; founder, Ontological-Hysteric Theater

Mistakes, errors, false starts—accept them all. The basis of creativity.

My reference point (as a playwright, not a scientist) was Keats's notion of negative capability (from his letters). Being able to exist with lucidity and calm amid uncertainty, mystery, and doubt, without "irritable [and always premature] reaching after fact and reason."

This toolkit notion of negative capability is a profound *therapy* for all manner of ills—intellectual, psychological, spiritual, and political. I reflect it (amplify it) with Emerson's notion that "Art [any intellectual activity?] is [best thought of as but] the *path* of the creator *to his work*."

Bumpy, twisting roads. (New York City is about to repave my cobblestoned street with smooth asphalt. Evil bureaucrats and tunnel-visioned "scientists"—fast cars and more tacky upscale stores in SoHo.)

Wow! I'll bet my contribution is shorter than anyone else's. Is this my inadequacy *or an important toolkit item heretofore overlooked*?

DEPTH

TOR NØRRETRANDERS
Science writer; consultant; lecturer; author, The Generous Man: How Helping Others Is the Sexiest Thing You Can Do

Depth is what you do not see immediately at the surface of things. Depth is what is below that surface: water below the surface of a lake, the rich life of soil, the spectacular line of reasoning behind a simple statement.

Depth is a straightforward aspect of the physical world. Gravity stacks stuff, and not everything can be at the top. Below there is more, and you can dig for it.

Depth acquired a particular meaning with the rise of complexity science a quarter of a century ago: What is characteristic of something complex? Orderly things, such as crystals, are not complex; they are simple. Messy things, such as a pile of litter, are difficult to describe; they hold a lot of information. Information is a measure of how difficult something is to describe. Disorder has a high information content and order has a low one. All the interesting stuff in life is in between: living creatures, thoughts, and conversations. Not a lot of information, but not a little, either. So information content does not lead us to what is interesting or complex. The marker is, rather, the information that is not there but was somehow involved in creating the object of interest. The history of the object is more relevant than the object itself, if we want to pinpoint what is interesting to us.

It is not the informational surface of the thing but its informational depth that attracts our curiosity. It took a lot to bring it here, before our eyes. It is not what is there but what used to be there that matters. Depth is about that.

The concept of depth in complexity science has been expressed in different ways: You can talk about the amount of physical information involved in bringing about something (the thermodynamic depth) or the amount of computation it took to arrive at a result (the logical depth). Both express the notion that the process behind is more important than the eventual product.

This idea can also be applied to human communication.

When you say "I do" at a wedding, it (one hopes) represents a huge amount of conversation, coexistence, and fun you've had with that other person. And a lot of reflection upon it. There is not a lot of information in the "I do" (one bit, actually), but the statement has depth.

Most conversational statements have some kind of depth. There is more than meets the ear, something that happened between the ears of the person talking before the statement was made. When you understand the statement, the meaning of what is being said, you "dig it," you get the depth, what is below and behind. What is not said but meant—the exformation content, information processed and thrown away before the actual production of explicit information.

2 + 2 = 4. This is a simple computation. The result, 4, holds less information than the problem, 2 + 2 (essentially because the problem could also have been 3 + 1 and yet the result would still be 4). Computation is wonderful as a method for throwing away information, getting rid of it. You do computations to ignore all the details, to get an overview, an abstraction, a result.

What you want is a way to distinguish between a very deep "I do" and a very shallow one: Did the guy actually think about what he said? Was the result "4" actually the result of a meaningful calculation? Is there, in fact, water below that surface? Does it have depth?

Most human interaction is about that question: Is this a bluff or for real? Is there sincere depth in the affection? Does the result stem from intense analysis or is it just an estimate? Is there anything between the lines?

Signaling is all about this question: fake or depth? In biology, the past few decades have seen the rise of studies on how animals prove to one another that there is depth behind a signal. The handicap principle of sexual selection is about a way to prove that your signal has depth: If a peacock has long, spectacular feathers, it proves that it can survive its predators despite its fancy plumage, which represents a disadvantage, a handicap. Hence, the peahen can know that the individual displaying the huge tail is a strong one or else it could not have survived with that extreme tail.

Among humans, you have what economists call costly signals, ways to show that you have something of value. The phenomenon of conspicuous consumption was observed by sociologist Thorstein Veblen in 1899: If you want to prove you have a lot of money, you have to waste it—that is, use it in a way that is absurd and idiotic, because only the rich can do so. But do it conspicuously, so that other people will know. Waste is a costly signal of the depth of a pile of money. Handicaps, costly signals, intense eye contact, and rhetorical gestures are all about proving that what seems so simple really has a lot of depth.

That is also the point with abstractions: We want them to be shorthand for a lot of information that was digested in the process leading to the use of the abstraction but not present when we use it. Such abstractions have depth. We love them. Other abstractions have no depth. They are shallow, just used to impress the other guy. They do not help us. We hate them.

Intellectual life is very much about the ability to distinguish between the shallow and the deep abstractions. You need to know if there is any depth before you make that headlong dive.

TEMPERAMENT DIMENSIONS

HELEN FISHER

Research professor, Department of Anthropology, Rutgers University; author, Why Him? Why Her?: How to Find and Keep Lasting Love

"I am large, I contain multitudes," wrote Walt Whitman. I have never met two people who were alike. I am an identical twin, and even we are not alike. Every individual has a distinct personality, a different cluster of thoughts and feelings that color all their actions. But there are patterns to personality: People express different styles of thinking and behaving—what psychologists call "temperament dimensions." I offer this concept of temperament dimensions as a useful new member of our cognitive toolkit.

Personality is composed of two fundamentally different types of traits, those of "character" and those of "temperament." Your character traits stem from your experiences. Your childhood games, your family's interests and values, how people in your community express love and hate, what relatives and friends regard as courteous or perilous, how those around you worship, what they sing, when they laugh, how they make a living and relax: Innumerable cultural forces build your unique set of character traits. The balance of your personality is your temperament, all the biologically based tendencies that contribute to your consistent patterns of feeling, thinking, and behaving. As the Spanish philosopher José Ortega y Gasset put it, "I am I, plus my circumstances." Temperament is the "I am I," the foundation of who you are.

Some 40 percent to 60 percent of the observed variance in personality is due to traits of temperament. They are heritable, relatively stable across the life course, and linked to specific gene pathways and/or hormone or neurotransmitter systems. Moreover, our temperament traits congregate in constellations, each aggregation associated with one of four broad, interrelated yet distinct brain systems: those associated with dopamine, serotonin, testosterone, and estrogen/oxytocin. Each constellation of temperament traits constitutes a distinct temperament dimension.

For example, specific alleles in the dopamine system have been linked with exploratory behavior, thrill, experience- and adventure-seeking, susceptibility to boredom, and lack of inhibition. Enthusiasm has been coupled with variations in the dopamine system, as have lack of introspection, increased energy and motivation, physical and intellectual exploration, cognitive flexibility, curiosity, idea generation, and verbal and nonlinguistic creativity.

The suite of traits associated with the serotonin system includes sociability, lower levels of anxiety, higher scores on scales of extroversion, and lower scores on a scale of "No Close Friends," as well as positive mood, religiosity, conformity, orderliness, conscientiousness, concrete thinking, self-control, sustained attention, low novelty-seeking, and figural and numeric creativity.

Heightened attention to detail, intensified focus, and narrow interests are some of the traits linked with prenatal testosterone expression. But testosterone activity is also associated with emotional containment, emotional flooding (particularly rage), social dominance and aggressiveness, less social sensitivity, and heightened spatial and mathematical acuity.

Last, the constellation of traits associated with the estrogen and related oxytocin system include verbal fluency and other language skills, empathy, nurturing, the drive to make social attachments and other prosocial aptitudes, contextual thinking, imagination, and mental flexibility.

We are each a different mix of these four broad temperament dimensions. But we do have distinct personalities. People are malleable, of course, but we are not blank slates upon which the environment inscribes personality. A curious child tends to remain curious, although what he or she is curious about changes with maturity. Stubborn people remain obstinate, orderly people remain punctilious, and agreeable men and women tend to remain amenable.

We are capable of acting "out of character," but doing so is tiring. People are biologically inclined to think and act in specific patterns—temperament dimensions. But why would this concept of temperament dimensions be useful in our human cognitive toolkit? Because we are social creatures, and a deeper understanding of who we (and others) are can provide a valuable tool for understanding, pleasing, cajoling, reprimanding, rewarding, and loving others—from friends and relatives to world leaders. It's also practical.

Take hiring. Those expressive of the novelty-seeking temperament dimension are unlikely to do their best in a job requiring rigid routines and schedules. Biologically cautious individuals are not likely to be comfortable in high-risk posts. Decisive, tough-minded high-testosterone types are not well suited to work with those who can't get to the point and decide quickly. And those predominantly of the compassionate, nurturing, high-estrogen temperament dimension are not likely to excel at occupations that require them to be ruthless.

Managers might form corporate boards containing all four broad types. Colleges might place freshmen with roommates of a similar temperament rather than similarity of background. Perhaps business teams, sports teams, political teams, and teacher-student teams would operate more effectively if they were either more "like-minded" or more varied in their cognitive skills. And certainly we could communicate with our children, lovers, colleagues, and friends more effectively. We are not puppets on a string of DNA. Those biologically susceptible to alcoholism, for example, often give up drinking. The more we come to understand our biology, the more we will appreciate how our culture molds it.

THE PERSONALITY/INSANITY CONTINUUM

GEOFFREY MILLER

Evolutionary psychologist; University of New Mexico; author, Spent: Sex, Evolution, and Consumer Behavior

We like to draw clear lines between normal and abnormal behavior. It's reassuring, for those who think they're normal. But it's not accurate. Psychology, psychiatry, and behavioral genetics are converging to show that there's no clear line between "normal variation" in human personality traits and "abnormal" mental illnesses. Our instinctive way of thinking about insanity—our intuitive psychiatry—is dead wrong.

To understand insanity, we have to understand personality. There's a scientific consensus that personality traits can be well described by five main dimensions of variation. These "Big Five" personality traits are called openness, conscientiousness, extroversion, agreeableness, and emotional stability. The Big Five are all normally distributed in a bell curve, statistically independent of one another, genetically heritable, stable across the life course, unconsciously judged when choosing mates or friends, and found in other species, such as chimpanzees. They predict a wide range of behaviors in school, work, marriage, parenting, crime, economics, and politics.

Mental disorders are often associated with maladaptive extremes of the Big Five traits. Overconscientiousness predicts obsessive-compulsive disorder, whereas low conscientiousness predicts drug addiction and other "impulse control" disorders. Low emotional stability predicts depression, anxiety, bipolar, borderline, and histrionic disorders. Low extroversion predicts avoidant and schizoid personality disorders. Low agreeableness predicts psychopathy and paranoid personality disorder. High openness is on a continuum with schizotypy and schizophrenia. Twin studies show that these links between personality traits and mental illnesses exist not just at the behavioral level but also at the genetic level. And parents who are somewhat extreme on a personality trait are much more likely to have a child with the associated mental illness.

One implication is that the "insane" are often just a bit more extreme in their personalities than whatever promotes success or contentment in modern societies—or more extreme than we're comfortable with. A less palatable implication is that we're all insane to some degree. All living humans have many mental disorders, mostly minor but some major, and these include not just classic psychiatric disorders like depression and schizophrenia but also diverse forms of stupidity, irrationality, immorality, impulsiveness, and alienation. As the new field of positive psychology acknowledges, we are all very far from optimal mental health, and we are all more or less crazy in many ways. Yet traditional psychiatry, like human intuition, resists calling anything a disorder if its prevalence is higher than about 10 percent.

The personality/insanity continuum is important in mental health policy and care. There are angry and unresolved debates over how to revise the fifth edition of psychiatry's core reference work, the *Diagnostic and Statistical Manual of Mental Disorders* (DSM-5), to be published in 2013. One problem is that American psychiatrists dominate the DSM-5 debates, and the American health insurance system demands discrete diagnoses of mental illnesses before patients are covered for psychiatric medications and therapies. Also, the U.S. Food and Drug Administration approves psychiatric medications only for discrete mental illnesses. These insurance and drug-approval issues push for definitions of mental illnesses to be artificially extreme, mutually exclusive, and based on simplistic checklists of symptoms. Insurers also want to save money, so they push for common personality variants—shyness, laziness, irritability, conservatism—not to be classed as illnesses worthy of care. But the science doesn't fit the insurance system's imperatives. It remains to be seen whether DSM-5 is written for the convenience of American insurers and FDA officials or for international scientific accuracy.

Psychologists have shown that in many domains our instinctive intuitions are fallible (though often adaptive). Our intuitive physics—ordinary concepts of time, space, gravity, and impetus—can't be reconciled with relativity, quantum mechanics, or cosmology. Our intuitive biology—ideas of species essences and teleological functions—can't be reconciled with evolution, population genetics, or adaptationism. Our intuitive morality—self-deceptive, nepotistic, clannish, anthropocentric, and punitive—can't be reconciled with any consistent set of moral values, whether Aristotelian, Kantian, or utilitarian. Apparently our intuitive psychiatry has similar limits. The sooner we learn those limits, the better we'll be able to help people with serious mental illnesses, and the more humble we'll be about our own mental health.

ARISE

JOEL GOLD
Psychiatrist; clinical assistant professor of psychiatry, NYU Langone Medical Center

ARISE, or Adaptive Regression In the Service of the Ego, is a psychoanalytic concept recognized for decades but little appreciated today. It is one of the ego functions, which, depending on whom you ask, may number anywhere from a handful to several dozen. They include reality testing, stimulus regulation, defensive function, and synthetic integration. For simplicity, we can equate the ego with the self (though ARISS doesn't quite roll off the tongue).

In most fields, including psychiatry, regression is not considered a good thing. Regression implies a return to an earlier and inferior state of being and functioning. But the key here is not the regression but rather whether the regression is maladaptive or adaptive.

There are numerous vital experiences that cannot be achieved without adaptive regression: The creation and appreciation of art, music, literature, and food; the ability to sleep; sexual fulfillment; falling in love; and, yes, the ability to free-associate and tolerate psychoanalysis or psychodynamic therapy without getting worse. Perhaps the most important element in adaptive regression is the ability to fantasize, to daydream. The person who has access to his unconscious processes and mines them without getting mired in them can try new approaches, can begin to see things in new ways, and, perhaps, can achieve mastery of his pursuits.

In a word: Relax.

It was ARISE that allowed Friedrich August Kekulé to use a daydream about a snake eating its tail as inspiration for his formulation of the structure of the benzene ring. It's what allowed Richard Feynman to simply drop an O-ring into a glass of ice water, show that when cold the ring loses pliability, and thereby explain the cause of the space shuttle *Challenger* disaster. Sometimes it takes a genius to see that a fifth-grade science experiment is all that is needed to solve a problem.

In another word: Play.

Sometimes in order to progress, you need to regress. Sometimes you just have to let go and ARISE.

SYSTEMIC EQUILIBRIUM

MATTHEW RITCHIE
Artist

The second law of thermodynamics, the so-called arrow of time, popularly associated with entropy (and by association, death), is the most widely misunderstood shorthand abstraction in human society today. We need to fix this.

The second law states that, over time, a closed system will become more homogeneous, eventually reaching systemic equilibrium. It is not a question of *whether* a system will reach equilibrium; it is a question only of *when* a system will reach equilibrium.

Living on a single planet, we are all participants in a single physical system that has only one direction—toward systemic equilibrium. The logical consequences are obvious; our environmental, industrial, and political systems (even our intellectual and theological systems) will become more homogeneous over time. It's already started. The physical resources available to every person on Earth, including air, food, and water, have already been significantly degraded by the high burn rate of industrialization, just as the intellectual resources available to every person on Earth have already been significantly increased by the high distribution rate of globalization.

Human societies are already far more similar than ever before (does anyone really miss dynastic worship?), and it would be very tempting to imagine that a modern democracy based on equal rights and opportunities is the system in equilibrium. That seems unlikely, given our current energy footprint. More likely, if the total system energy is depleted too fast, is that modern democracies will be compromised if the system crashes to its lowest equilibrium too quickly for socially equitable evolution.

Our one real opportunity is to use the certain knowledge of ever-increasing systemic equilibrium to build a model for an equitable and sustainable future. The mass distribution of knowledge and access to information through the World Wide Web is our civilization's signal achievement. Societies that adopt innovative, predictive, and adaptive models designed around a significant, ongoing redistribution of global resources will be most likely to survive in the future.

But since we are biologically and socially programmed to avoid discussing entropy (death), we reflexively avoid the subject of systemic changes to our way of life, both as a society and individuals. We think it's a bummer. Instead of examining the real problems, we consume apocalyptic fantasies as "entertainment" and deride our leaders for their impotence. We really need to fix this.

Unfortunately, even facing this basic concept is an uphill battle today. In earlier, expansionist phases of society, various metaphorical engines such as "progress" and "destiny" allowed the metaphorical arrow to supplant the previously (admittedly spirit-crushing) wheel of time. Intellectual positions that supported scientific experimentation and causality were tolerated, even endorsed, as long as they contributed to the arrow's cultural momentum. But in a more crowded and contested world, the limits of projected national power and consumption control have become more obvious. Resurgent strands of populism, radicalism, and magical thinking have found mass appeal in their rejection of many rational concepts. But perhaps most significant is the rejection of undisputed physical laws.

The practical effect of this denial of the relationship between the global economy and the climate-change debate (for example) is obvious. Advocates propose continuous "good" (green) growth, while denialists propose continuous "bad" (brown) growth. Both sides are more interested in backing winners and losers in a future economic environment predicated on the continuation of today's systems than in accepting the physical inevitability of increasing systemic equilibrium in any scenario.

Of course, any system can temporarily cheat entropy. Hotter particles (or societies) can "steal" the stored energy of colder (or weaker) ones, for a while. But in the end, the rate at which the total energy is burned and redistributed will still determine the speed at which the planetary system will reach its true systemic equilibrium. Whether we extend the lifetime of our local "heat" through war or improved window insulation is the stuff of politics. But even if in reality we can't beat the house, it's worth a try, isn't it?

PROJECTIVE THINKING

LINDA STONE

High-tech industry consultant; former executive, Apple Computer and Microsoft Corporation

Barbara McClintock was ignored and ridiculed by the scientific community for thirty-two years before winning the 1983 Nobel Prize in physiology or medicine for discovering "jumping genes." During the years of hostile treatment by her peers, McClintock didn't publish, preferring to avoid the rejection of the scientific community. Stanley Prusiner faced significant criticism from his colleagues until his prion theory was confirmed. He, too, went on to win a Nobel Prize, in 1997.

Barry Marshall challenged the medical "fact" that stomach ulcers were caused by acid and stress and presented evidence that bacterial infection by *H. pylori* is the cause. Marshall noted in an 1998 interview that "Everyone was against me."

Progress in medicine was delayed while these "projective thinkers" persisted, albeit on a slower and lonelier course.

"Projective thinking" is a term coined by Edward de Bono to describe generative rather than reactive thinking. McClintock, Prusiner, and Marshall offered projective thinking, suspending their disbelief regarding scientific views accepted at the time.

Articulate, intelligent individuals can skillfully construct a convincing case to argue almost any point of view. This critical, reactive use of intelligence narrows our vision. In contrast, projective thinking is expansive, "open-ended," and speculative, requiring the thinker to create the context, concepts, and the objectives.

Twenty years of studying maize created a context within which McClintock could speculate. With her extensive knowledge and keen powers of observation, she deduced the significance of the changing color patterns of maize seed. This led her to propose the concept of gene regulation, which challenged the theory of the genome as a static set of instructions passed from one generation to the next. The work McClintock first reported in 1950, the result of projective thinking, extensive research, persistence, and a willingness to suspend disbelief, wasn't understood or accepted until many years later.

Everything we know, our strongly held beliefs, and in some cases even what we consider to be "factual," creates the lens through which we see and experience the world and can contribute to a critical, reactive orientation. This can serve us well. (Fire is hot; it can burn if touched.) It can also compromise our ability to observe and to think in an expansive, generative way.

When we cling rigidly to our constructs, as McClintock's peers did, we can be blinded to what's right in front of us. Can we support a scientific rigor that embraces generative thinking and suspension of disbelief? Sometimes science fiction does become scientific discovery.

ANOMALIES AND PARADIGMS

V. S. RAMACHANDRAN
Neuroscientist; director, Center for Brain and Cognition, University of California–San Diego; author, The Tell-Tale Brain *and* Phantoms in the Brain

Do you need language for sophisticated thinking, or do words merely facilitate thought? This question goes back to a debate between two Victorian scientists, Max Mueller and Francis Galton.

A word that has made it into the common vocabulary of both science and pop culture is "paradigm"—and the converse, "anomaly"—the former having been introduced by the historian of science Thomas Kuhn. "Paradigm" is now widely used and misused both in science and in other disciplines, almost to the point where the original meaning is starting to be diluted. (This often happens to "memes" of human language and culture, which don't enjoy the lawful, particulate transmission of genes.) The word "paradigm" is now often used inappropriately, especially in the United States, to mean any experimental procedure—such as "the Stroop paradigm" or "a reaction-time paradigm" or "the fMRI paradigm."

However, its appropriate use has shaped our culture in significant ways, even influencing the way scientists work and think. A more prevalent associated word is "skepticism," originating from the name of a Greek school of philosophy . This is used even more frequently and loosely than "anomaly" and "paradigm shift."

One can speak of reigning paradigms—what Kuhn calls normal science and what I cynically refer to as a mutual-admiration club trapped in a cul-de-sac of specialization. The club usually has its pope(s), hierarchical priesthood, acolytes, and a set of guiding assumptions and accepted norms zealously guarded with almost religious fervor. (Its members also fund one another, and review one another's papers and grants, and give one another awards.)

This isn't entirely useless; it's "normal science" that grows by progressive accretion, employing the bricklayers rather than the architects of science. If a new experimental observation (e.g., bacterial transformation; ulcers cured by antibiotics) threatens to topple the edifice, it's called an anomaly, and the typical reaction of those who practice normal science is to ignore it or brush it under the carpet—a form of psychological denial surprisingly common among my colleagues.

This is not an unhealthy reaction, since most anomalies turn out to be false alarms; the baseline probability of their survival as real anomalies is small, and whole careers have been wasted pursuing them (think polywater and cold fusion). Yet even such false anomalies serve the useful purpose of jolting scientists from their slumber by calling into question the basic axioms that drive their particular area of science. Conformist science feels cozy, given the gregarious nature of humans, and anomalies force periodic reality checks even if the anomaly turns out to be flawed.

More important, though, are genuine anomalies that emerge every now and then, legitimately challenging the status quo, forcing paradigm shifts, and leading to scientific revolutions. Conversely, premature skepticism toward anomalies can lead to stagnation of science. One needs to be skeptical of anomalies but equally skeptical of the status quo if science is to progress.

I see an analogy between the process of science and of evolution by natural selection. For evolution, too, is characterized by periods of stasis (= normal science), punctuated by brief periods of accelerated change (= paradigm shifts), based on mutations (= anomalies), most of which are lethal (false theories) but some of which lead to the budding-off of new species and phylogenetic trends (= paradigm shifts).

Since most anomalies are false alarms (spoon-bending, telepathy, homeopathy), one can waste a lifetime pursuing them. So how does one decide which anomalies to invest in? Obviously one can do so by trial and error, but that can be tedious and time-consuming.

Let's take four well-known examples: (1) continental drift, (2) bacterial transformation, (3) cold fusion, and (4) telepathy. All of these were anomalies when they arose, because they didn't fit the big picture of normal science at the time. The evidence that all the continents broke off and drifted away from a giant supercontinent was staring people in the face, as Wegener noted in the early twentieth century. Coastlines coincided almost perfectly; certain fossils found on the east coast of Brazil were exactly the same as the ones on the west coast of Africa, etc. Yet it took fifty years for the idea to be accepted by the skeptics.

Anomaly (2) was observed by Fred Griffith, decades before DNA and the genetic code. He found that if you inject a heat-treated, dead, virulent species of bacteria (*pneumococcus S*) into a rat previously infected with a nonvirulent species (*pneumococcus R*), then species R became *transformed* into species S, thereby killing the rat. About fifteen years later, Oswald Avery found that you can even do this in a test tube; dead S would transform live R into live S if the two were simply incubated together; moreover, the change was heritable. Even the juice from S did the trick, leading Avery to suspect that a chemical substance in the juice—DNA—might be the carrier of heredity. Others replicated this. It was almost like saying, "Put a dead lion and eleven pigs into a room and a dozen live lions emerge," yet the discovery was largely ignored for years. Until Watson and Crick deciphered the mechanism of transformation.

The third anomaly—telepathy—is almost certainly a false alarm.

You will see a general rule of thumb emerging here. Anomalies (1) and (2) were not ignored because of lack of empirical evidence.

Even a schoolchild can see the fit between continental coastlines or the similarity of fossils. (1) was ignored solely because it didn't fit the big picture—the notion of *terra firma*, or a solid, immovable Earth—and there was no conceivable mechanism that would allow continents to drift, until plate tectonics was discovered. Likewise (2) was repeatedly confirmed but ignored because it challenged the fundamental doctrine of biology—the stability of species. But notice that the third, telepathy, was rejected for two reasons: first, because it didn't fit the big picture; and second, because it was hard to replicate. This gives us the recipe we are looking for: focus on anomalies that have survived repeated attempts to disprove experimentally but are ignored by the establishment *solely* because you can't think of a mechanism. But don't waste time on ones that have not been empirically confirmed despite repeated attempts (or ones for which the effect becomes smaller with each attempt—a red flag!)

Words themselves are paradigms, or stable "species" of sorts, that evolve gradually with progressively accumulating penumbras of meaning or sometimes mutate into new words to denote new concepts. These can then consolidate into chunks with "handles" (names) for juggling ideas around, generating novel combinations. As a behavioral neurologist, I am tempted to suggest that such crystallization of words, and juggling them, is unique to humans and occurs in brain areas in and near the left TPO (temporal-parietal-occipital junction). But that's pure speculation.

RECURSIVE STRUCTURE

DAVID GELERNTER
Computer scientist, Yale University; chief scientist, Mirror Worlds Technologies; author, Mirror Worlds

Recursive structure is a simple idea (or shorthand abstraction) with surprising applications beyond science.

A structure is recursive if the shape of the whole recurs in the shape of the parts: for example, a circle formed of welded links that are circles themselves. Each circular link might itself be made of smaller circles, and in principle you could have an unbounded nest of circles made of circles made of circles.

The idea of recursive structure came into its own with the advent of computer science (that is, software science) in the 1950s. The hardest problem in software is controlling the tendency of software systems to grow incomprehensibly complex. Recursive structure helps convert impenetrable software rain forests into French gardens—still (potentially) vast and complicated but much easier to traverse and understand than a jungle.

Benoit Mandelbrot famously recognized that some parts of nature show recursive structure of a sort: A typical coastline shows the same shape or pattern whether you look from six inches or sixty feet or six miles away.

But it also happens that recursive structure is fundamental to the history of architecture, especially to the Gothic, Renaissance, and Baroque architecture of Europe—covering roughly the five hundred years between the thirteenth and eighteenth centuries. The strange case of "recursive architecture" shows us the damage one missing idea can create. It suggests also how hard it is to talk across the cultural Berlin Wall that separates science and art. And the recurrence of this phenomenon in art and nature underlines an important aspect of the human sense of beauty.

The reuse of one basic shape on several scales is fundamental to Medieval architecture. But, lacking the idea (and the term) "recursive structure," art historians are forced to improvise ad-hoc descriptions each time they need one. This hodgepodge of improvised descriptions makes it hard, in turn, to grasp how widespread recursive structure really is. And naturally, historians of post-Medieval art invent their own descriptions—thus obfuscating a fascinating connection between two mutually alien aesthetic worlds.

For example: One of the most important aspects of mature Gothic design is tracery—the thin, curvy, carved stone partitions that divide one window into many smaller panes. Recursion is basic to the art of tracery.

Tracery was invented at the cathedral of Reims circa 1220 and used soon after at the cathedral of Amiens. (Along with Chartres, these two spectacular and profound buildings define the High Gothic style.) To move from the characteristic tracery design of Reims to that of Amiens, just add recursion. At Reims, the basic design is a pointed arch with a circle inside; the circle is supported on two smaller arches. At Amiens, the basic design is the same—except that now the window recurs in miniature inside each smaller arch. (Inside each smaller arch is a still smaller circle supported on still smaller arches.)

In the great east window at Lincoln Cathedral, the recursive nest goes one step deeper. This window is a pointed arch with a circle inside; the circle is supported on two smaller arches—much like Amiens. Within each smaller arch is a circle supported on two still smaller arches. Within each still smaller arch, a circle is supported on even smaller arches.

There are other recursive structures throughout Medieval art. Jean Bony and Erwin Panofsky were two eminent twentieth-century art historians. Naturally, they both noticed recursive structure. But neither man understood *the idea in itself.* And so, instead of writing that the windows of Saint-Denis show recursive structure, Bony said that they are "composed of a series of similar forms progressively subdivided in increasing numbers and decreasing sizes." Describing the same phenomenon in a different building, Panofsky writes of the "principle of progressive divisibility (or, to look at it the other way, multiplicability)." Panofsky's "principle of progressive divisiblity" is a fuzzy, roundabout way of saying "recursive structure."

Louis Grodecki noticed the same phenomenon—a chapel containing a display platform shaped like the chapel in miniature, holding a shrine shaped like the chapel in extra-miniature. And he wrote: "This is a common principle of Gothic art." But he doesn't say *what* the principle is; he doesn't describe it *in general* or give it a name. William Worringer, too, had noticed recursive structure. He described Gothic design as "a world which repeats in miniature, but with the same means, the expression of the whole."

So each historian makes up his own name and description for the same basic idea—which makes it hard to notice that all four descriptions actually describe the same thing. *Recursive structure is a basic principle of Medieval design*; but this simple statement is hard to say, or even think, if we don't know what "recursive structure" is.

If the literature makes it hard to grasp the importance of recursive structure in Medieval art, it's even harder to notice that exactly the same principle recurs in the radically different world of Italian Renaissance design.

George Hersey wrote astutely of Bramante's design (ca. 1500) for St. Peter's in the Vatican that it consists of "a single macrochapel . . . , four sets of what I will call maxichapels, sixteen minichapels, and thirty-two microchapels." "The principle [he explains] is that of Chinese boxes—or, for that matter, fractals." If only he had been able to say that "recursive structure is fundamental to Bramante's thought," the whole discussion would have been simpler and clearer—and an intriguing connection between Medieval and Renaissance design would have been obvious.

Using instead of ignoring the idea of recursive structure would have had other advantages, too. It helps us understand the connections between art and technology, helps us see the aesthetic principles that guide the best engineers and technologists and the ideas of clarity and elegance that underlie every kind of successful design. These ideas have practical implications. For one, technologists must study and understand elegance and beauty as design goals; any serious technology education must include art history. And we reflect, also, on the connection between great art and great technology on the one hand and natural science on the other.

But without the right intellectual tool for the job, new instances of recursive structure make the world more complicated instead of simpler and more beautiful.

DESIGNING YOUR MIND

DON TAPSCOTT
Business strategist; chairman, Moxie Insight; adjunct professor, Rotman School of Management, University of Toronto; author, Grown Up Digital: How the Net Generation Is Changing Your World; *coauthor (with Anthony D. Williams),* Macrowikinomics: Rebooting Business and the World

Given recent research about brain plasticity and the dangers of cognitive load, the most powerful tool in our cognitive arsenal may well be design. Specifically, we can use design principles and discipline to shape our minds. This is different from acquiring knowledge. It's about designing how each of us thinks, remembers, and communicates—appropriately and effectively for the digital age.

Today's popular hand-wringing about the digital age's effects on cognition has some merit. But rather than predicting a dire future, perhaps we should be trying to achieve a new one. New neuroscience discoveries give hope. We know that brains are malleable and can change depending on how they are used. The well-known study of London taxi drivers showed that a certain region in the brain involved in memory formation was physically larger than in non-taxi-driving individuals of a similar age. This effect did not extend to London bus drivers, supporting the conclusion that the requirement of London's taxi drivers to memorize the multitude of London streets drove structural brain changes in the hippocampus.

Results from studies like these support the notion that even among adults the persistent, concentrated use of one neighborhood of the brain really can increase its size and presumably its capacity. Not only does intense use change adult brain regional structure and function but temporary training and perhaps even mere mental rehearsal seem to have an effect as well. A series of studies showed that one can improve tactile (Braille character) discrimination among seeing people who are blindfolded. Brain scans revealed that participants' visual cortex responsiveness was heightened to auditory and tactile sensory input after only five days of blindfolding for over an hour each time.

The existence of lifelong neuroplasticity is no longer in doubt. The brain runs on a "use it or lose it" motto. So could we use it to build it right? Why don't we use the demands of our information-rich, multistimuli, fast-paced, multitasking digital existence to expand our cognitive capability? Psychiatrist Dr. Stan Kutcher, an expert on adolescent mental health who has studied the effect of digital technology on brain development, says we probably can: "There is emerging evidence suggesting that exposure to new technologies may push the Net Generation [teenagers and young adults] brain past conventional capacity limitations."

When the straight-A student is doing her homework at the same time as five other things online, she is not actually multitasking. Instead, she has developed better active working memory and better switching abilities. I can't read my e-mail and listen to iTunes at the same time, but she can. Her brain has been wired to handle the demands of the digital age.

How could we use design thinking to change the way we think? Good design typically begins with some principles and functional objectives. You might wish to perceive and absorb information effectively, concentrate, remember, infer meaning, be creative, write, speak, and communicate well, and enjoy important collaborations and human relationships. How could you design your use of (or abstinence from) media to achieve these goals?

Something as old-school as a speed-reading course could increase your input capacity without undermining comprehension. If it made sense in Evelyn Wood's day, it is doubly important now, and we've learned a lot since then about how to read effectively.

Feeling distracted? The simple discipline of reading a few full articles per day rather than just the headlines and summaries could strengthen attention.

Want to be a surgeon? Become a gamer, or rehearse while on the subway. Rehearsal can produce changes in the motor cortex as big as those induced by physical movement. In one study, a group of participants was asked to play a simple five-finger exercise on the piano while another group of participants was asked to think about playing the same tune in their heads using the same finger movements, one note at a time. Both groups showed a change in their motor cortex, with differences among the group who mentally rehearsed the tune as great as those who did so physically.

Losing retention? Decide how far you want to apply Albert Einstein's law of memory. When asked why he went to the phone book to get his number, he replied that he memorized only those things he couldn't look up. There's a lot to remember these days. Between the dawn of civilization and 2003, there were five exabytes of data collected (an exabyte equals 1 quintillion bytes). Today five exabytes of data gets collected every two days! Soon there will be five exabytes every few minutes. Humans have a finite memory capacity. Can you develop criteria for which will be inboard and which outboard?

Or want to strengthen your working memory and ability to multitask? Try reverse mentoring—learning with your teenager. This is the first time in history when children are authorities about something important, and the successful ones are pioneers of a new paradigm in thinking. Extensive research shows that people can improve cognitive function and brain efficiency through simple lifestyle changes, such as incorporating memory exercises into their daily routine.

Why don't schools and universities teach design for thinking? We teach physical fitness, but rather than brain fitness, we emphasize cramming young heads with information and testing their recall. Why not courses that emphasize designing a great brain?

Does this modest proposal raise the specter of "designer minds"? I don't think so. The design industry is something done to us. I'm proposing that we each become designers.

FREE JAZZ

ANDRIAN KREYE
Editor, The Feuilleton (arts and essays) *of the German daily* Sueddeutsche Zeitung, *Munich*

It's always worth taking a few cues from the mid-twentieth-century avant-garde. When it comes to improving your cognitive toolkit, free jazz is perfect. It is a highly evolved new take on an art that has—at least, in the West—been framed by a strict set of twelve notes played in accurate fractions of bars. It is also the pinnacle of a genre that began with the blues, just a half century before Ornette Coleman assembled his infamous double quartet in the A&R Studio in New York City one December day in 1960. In science terms, that would mean an evolutionary leap from elementary-school math to game theory and fuzzy logic in a mere fifty years.

If you really want to appreciate the mental prowess of free-jazz players and composers, you should start just one step behind. A half year before Coleman's free-jazz session let loose the improvisational genius of eight of the best musicians of their time, John Coltrane recorded what is still considered the most sophisticated jazz solo ever—his tour de force through the rapid chord progressions of his composition "Giant Steps." The film student Daniel Cohen has recently animated the notation for Coltrane's solo in a YouTube video. You don't have to be able to read music to grasp the intellectual firepower of Coltrane. After the deceivingly simple main theme, the notes start to race up and down the five lines of the stave in dizzying speeds and patterns. If you also take into consideration that Coltrane used to record unrehearsed music to keep it fresh, you know that he was endowed with a cognitive toolkit way beyond normal.

Now take these four minutes and forty-three seconds, multiply Coltrane's firepower by eight, stretch it into thirty-seven minutes, and deduct all traditional musical structures, like chord progressions or time. The 1960 session that gave the genre its name in the first place foreshadowed not just the radical freedom the album's title, *Free Jazz: A Collective Improvisation by the Ornette Coleman Double Quartet*, implied. It was a precursor to a form of communication that has left linear conventions and entered the realm of multiple parallel interactions.

It is admittedly still hard to listen to the album. It is equally taxing to listen to recordings of Cecil Taylor, Pharoah Sanders, Sun Ra, Anthony Braxton, or Gunter Hampel. It has always been easier to understand the communication processes of this music in a live setting. One thing is a given—it is never anarchy, never was meant to be.

If you're able to play music and you manage to get yourself invited to a free-jazz session, you'll experience the incredible moment when all the musicians find what is considered "the pulse." It is a collective climax of creativity and communication that can leap to the audience and create an electrifying experience. It's hard to describe but might be comparable to the moment when the catalyst of a surfboard brings together the motor skills of the surfer's body and the forces of the ocean's swell, in those few seconds of synergy on top of a wave. It is a fusion of musical elements, though, that defies common musical theory.

Of course, there is a lot of free jazz that merely confirms prejudice. Or as vibraphonist/composer Hampel phrased it: "At one point it was just about being the loudest onstage." But the musicians mentioned above found new forms and structures, Ornette Coleman's music theory called Harmolodics being just one of them. In the perceived cacophony of their music, there is a multilayered clarity that can serve as a model for a cognitive toolkit for the twenty-first century. The ability to find cognitive, intellectual, and communication skills that work in parallel contexts rather than linear forms will be crucial. Just as free jazz abandoned harmonic structures to find new forms in polyrhythmic settings, one may have to enable oneself to work beyond proven cognitive patterns.

COLLECTIVE INTELLIGENCE

MATT RIDLEY

Science writer; founding chairman, International Centre for Life; author, The Rational Optimist: How Prosperity Evolves

Brilliant people, be they anthropologists, psychologists, or economists, assume that brilliance is the key to human achievement. They vote for the cleverest people to run governments, they ask the cleverest experts to devise plans for the economy, they credit the cleverest scientists with discoveries, and they speculate on how human intelligence evolved in the first place.

They are all barking up the wrong tree. The key to human achievement is not individual intelligence at all. The reason human beings dominate the planet is not because they have big brains: Neanderthals had big brains but were just another kind of predatory ape. Evolving a 1,200-cubic-centimeter brain and a lot of fancy software like language was necessary but not sufficient for civilization. The reason some economies work better than others is certainly not because they have cleverer people in charge, and the reason some places make great discoveries is not because the people there are smarter.

Human achievement is entirely a networking phenomenon. It is by putting brains together through the division of labor—through trade and specialization—that human society stumbled upon a way to raise the living standards, carrying capacity, technological virtuosity, and knowledge base of the species. We can see this in all sorts of phenomena: the correlation between technology and connected population size in Pacific islands; the collapse of technology in people who became isolated, like native Tasmanians; the success of trading city-states in Greece, Italy, Holland, and Southeast Asia; the creative consequences of trade.

Human achievement is based on collective intelligence—the nodes in the human neural network are people themselves. By each doing one thing and getting good at it, then sharing and combining the results through exchange, people become capable of doing things *they do not even understand*. As the economist Leonard Read observed in his essay "I, Pencil" (which I'd like everybody to read), no single person knows how to make even a pencil—the knowledge is distributed in society among many thousands of graphite miners, lumberjacks, designers, and factory workers.

That's why, as Friedrich Hayek observed, central planning never worked: The cleverest person is no match for the collective brain at working out how to distribute consumer goods. The idea of bottom-up collective intelligence, which Adam Smith understood and Charles Darwin echoed, and which Hayek expounded in his remarkable essay "The Use of Knowledge in Society," is one idea I wish everybody had in their cognitive toolkit.

RISK LITERACY

GERD GIGERENZER

Psychologist; director of the Center for Adaptive Behavior and Cognition at the Max Planck Institute for Human Development, Berlin; author, Gut Feelings

Literacy is the precondition for an informed citizenship in a participatory democracy. But knowing how to read and write is no longer enough. The breakneck speed of technological innovation has made risk literacy as indispensable in the twenty-first century as reading and writing were in the twentieth. Risk literacy is the ability to deal with uncertainties in an informed way.

Without it, people jeopardize their health and money and can be manipulated into experiencing unwarranted, even damaging, hopes and fears. Yet when considering how to deal with modern threats, policy makers rarely ever invoke the concept of risk literacy in the general public. To reduce the chances of another financial crisis, proposals called for stricter laws, smaller banks, reduced bonuses, lower leverage ratios, less short-termism, and other measures. But one crucial idea was missing: helping the public better understand financial risk. For instance, many of the NINJAs (No Income, No Job, No Assets) who lost everything but the shirts on their backs in the subprime crisis hadn't realized that their mortgages were variable, not fixed-rate.

Another serious problem that risk literacy can help solve is the exploding cost of health care. Tax hikes or rationed care are often presented as the only viable solutions. Yet by promoting health literacy in patients, better care can be had for less money. For instance, many parents are unaware that a million U.S. children have unnecessary CT scans annually, and that a full-body scan

can deliver a thousand times the radiation dose of a mammogram, resulting in an estimated twenty-nine thousand cancers per year.

I believe that the answer to modern crises is not simply more laws, more bureaucracy, or more money, but, first and foremost, more citizens who are risk-literate. This can be achieved by cultivating statistical thinking.

Simply stated, statistical thinking is the ability to understand and critically evaluate uncertainties and risks. Yet 76 percent of U.S. adults and 54 percent of Germans do not know how to express a 1 in 1,000 chance as a percentage (0.1 percent). Schools spend most of their time teaching children the mathematics of certainty—geometry, trigonometry—and little if any time on the mathematics of uncertainty. If taught at all, it is mostly in the form of coin and dice problems that tend to bore young students to death. But statistical thinking could be taught as the art of real-world problem solving—i.e., the risks of drinking, AIDS, pregnancy, skateboarding, and other dangerous things. Out of all mathematical disciplines, statistical thinking connects most directly to a teenager's world.

At the university level, law and medical students are rarely taught statistical thinking, even though they are pursuing professions whose very nature it is to deal with matters of uncertainty. U.S. judges and lawyers have been confused by DNA statistics; their British colleagues have drawn incorrect conclusions about the probability of recurring sudden infant death. Many doctors worldwide misunderstand the likelihood that a patient has cancer after a positive screening test, or can't critically evaluate new evidence presented in medical journals. Experts without risk-literacy skills are part of the problem rather than the solution.

Unlike basic literacy, risk literacy requires emotional rewiring—rejecting comforting paternalism and illusions of certainty and learning to take responsibility and to live with uncertainty.

Daring to know. But there is still a long way to go. Studies indicate that most patients want to believe in their doctor's omniscience and don't dare to ask for supporting evidence, yet nevertheless feel well-informed after consultations. Similarly, even after the banking crisis, many customers still blindly trust their financial advisors, jeopardizing their fortunes in a consultation that takes less time than they'd spend watching a football game. Many people cling to the belief that others can predict the future and pay fortune-tellers for illusory certainty. Every fall, renowned financial institutions forecast next year's Dow and dollar exchange rate, even though their track record is hardly better than chance. We pay $200 billion yearly to a forecasting industry that delivers mostly erroneous future predictions.

Educators and politicians alike should realize that risk literacy is a vital topic for the twenty-first century. Rather than being nudged into doing what experts believe is right, people should be encouraged and equipped to make informed decisions for themselves. Risk literacy should be taught beginning in elementary school. Let's dare to know—risks and responsibilities are chances to be taken, not avoided.

SCIENCE VERSUS THEATER

ROSS ANDERSON

Professor of security engineering, University of Cambridge Computer Laboratory; researcher in the economics and psychology of information security

Modern societies waste billions on protective measures whose real aim is to reassure rather than to reduce risk. Those of us who work in security engineering refer to this as "security theater," and there are examples all around us. We're searched going into buildings that no terrorist would attack. Social-network operators create the pretense of a small intimate group of "friends," in order to inveigle users into disclosing personal information that can be sold to advertisers. The users get not privacy but privacy theater. Environmental policy is a third example: Cutting carbon emissions would cost lots of money and votes, so governments go for gesture policies that are highly visible though their effect is negligible. Specialists know that most of the actions that governments claim will protect the security of the planet are just theater.

Theater thrives on uncertainty. Wherever risks are hard to measure or their consequences hard to predict, appearance can be easier to manage than reality. Reducing uncertainty and exposing gaps between appearance and reality are among the main missions of science.

Our traditional approach was the painstaking accumulation of knowledge that enables people to understand risks, options, and consequences. But theater is a deliberate construct rather than an accidental side effect of ignorance, so perhaps we need to become more sophisticated about theatrical mechanisms too. Science communicators need to become adept at disrupting the show, illuminating the dark corners of the stage, and making the masks visible for what they are.

THE BASE RATE

KEITH DEVLIN
Executive director, H-STAR Institute, Stanford University; author, The Unfinished Game: Pascal, Fermat, and the Seventeenth-Century Letter That Made the World Modern

The recent controversy about the potential dangers to health of the back-scatter radiation devices being introduced at the nation's airports and the intrusive patdowns offered as the only alternative by the TSA might well have been avoided, had citizens been aware of, and understood, the probabilistic notion of base rate.

Whenever a statistician wants to predict the likelihood of some event based on the available evidence, there are two main sources of information that have to be taken into account: (1) the evidence itself, for which a reliability figure has to be calculated; and (2) the likelihood of the event calculated purely in terms of relative incidence. The second figure here is the base rate. Since it is just a number, obtained by the seemingly dull process of counting, it frequently gets overlooked when there is new information, particularly if that new information is obtained by "experts" using expensive equipment. In cases where the event is dramatic and scary, like a terrorist attack on an airplane, failure to take account of the base rate can result in wasting massive amounts of effort and money trying to prevent something that is very unlikely.

For example, suppose that you undergo a medical test for a relatively rare cancer. The cancer has an incidence of 1 percent among the general population. (That is the base rate.) Extensive trials have shown that the reliability of the test is 79 percent. More precisely, although the test does not fail to detect the cancer when it is present, it gives a positive result in 21 percent of the cases where no cancer is present—what is known as a false positive. When you are tested, the test produces a positive diagnosis. The question is: What is the probability that you have the cancer?

If you're like most people, you'll assume that if the test has a reliability rate of nearly 80 percent, and you test positive, then the likelihood that you do indeed have the cancer is about 80 percent (i.e., the probability is approximately 0.8). Are you right?

The answer is no. You have focused on the test and its reliability and overlooked the base rate. Given the scenario just described, the likelihood that you have the cancer is a mere 4.6 percent (i.e., 0.046). That's right—there is a less than 5 percent chance that you have the cancer. Still a worrying possibility, of course. But hardly the scary 80 percent you thought at first.

In the case of the back-scatter radiation devices at the airports, the base rate for dying in a terrorist attack is lower than many other things we do every day without hesitation. In fact, according to some reports, it is about the same as the likelihood of getting cancer as a result of going through the device.

FINDEX

MARTI HEARST
Computer scientist, University of California–Berkeley, School of Information; author, Search User Interfaces

***Findex (n):* The degree to which a desired piece of information can be found online.**

We are the first humans in history to be able to form just about any question in our minds and know that very likely the answer can be before us in minutes, if not seconds. This omnipresent information abundance is a cognitive toolkit entirely in itself. The actuality of this continues to astonish me.

Although some have written about information overload, data smog, and the like, my view has always been the more information online the better, as long as good search tools are available. Sometimes this information is found by directed search using a Web search engine, sometimes by serendipity by following links, and sometimes by asking hundreds of people in your social network or hundreds of thousands of people on a question-answering Web site such as Answers.com, Quora, or Yahoo Answers.

I do not actually know of a real findability index, but tools in the field of information retrieval could be applied to develop one. One of the unsolved problems in the field is how to help the searcher to determine if the information simply is not available.

AN ASSERTION IS OFTEN AN EMPIRICAL QUESTION, SETTLED BY COLLECTING EVIDENCE

SUSAN FISKE

Eugene Higgins Professor of Psychology, Princeton University; author, Envy Up, Scorn Down: How Status Divides Us

The most important scientific concept is that an assertion is often an empirical question, settled by collecting evidence. The plural of anecdote is not data, and the plural of opinion is not facts. Quality peer-reviewed scientific evidence accumulates into knowledge. People's stories are stories, and fiction keeps us going. But science should settle policy.

SCIENTISTS SHOULD BE SCIENTISTS

GREGORY PAUL

Independent researcher; author, Dinosaurs of the Air: The Evolution and Loss of Flight in Dinosaurs and Birds

The archenemy of scientific thinking is conversation, as in typical human conversational discourse, much of which is BS. I have become rather fed up with talking to people. Seriously, it is something of a problem. Fact is, folks are prone to getting pet opinions into their heads and thinking they're true to the point of obstinacy, even when they have little or no idea of what they're talking about in the first place. We all do it. It's part of how the sloppy mind-generating piece of meat between our ears is prone to work. Humans may be the most rational beings on the planet these days—but that's not saying much, considering that the next most rational are chimpanzees.

Take creationism. Along with the global-climate issue and parental fear of vaccination, the fact that a big chunk of the American body politic denies evolutionary and paleontological science and actually thinks a god created humans in near historical times is causing scientists to wonder just what is wrong with the thinking of so many people. Mass creationism has been used as a classic example of mass antiscientific thinking by others responding to this question. But I am not going to focus so much on the usual problem of why creationism is popular but more on what many who promote science over creationism think they know about those who deny the reality of Darwin's theory.

A few years back, an anticreationist documentary came out titled *A Flock of Dodos*. Nicely done in many regards, the documentary scored some points against the antievolution crowd, but when it came to trying to explain why many Americans are repelled by evolution, it was way off base. The reason it was so wrong is that the creator of the film, Randy Olson, went to the wrong people to find out where the problem lies. A highlight of the picture featured a bunch of poker-playing Harvard evolutionary scientists gathered around a table to converse and opine on why the yahoos don't like the results of their research. This was a very bad mistake, for the simple reason that evolutionary scientists are truly knowledgeable only about their area of expertise, evolutionary science.

If you really want to know why regular folks think the way they do, then you go to the experts on that subject, sociologists. Because *A Flock of Dodos* never does that, its viewers never find out why creationism thrives in the age of science and what needs to be done to tame the pseudoscientific beast.

This is not an inconsequential problem. In the last decade, big strides have been made in understanding the psychosociology of popular creationism. Basically, it flourishes only in seriously dysfunctional societies, and the one sure way to suppress the errant belief is to run countries well enough so that the religion that creationism depends upon withers to minority status, dragging creationism down with it.

In other words, better societies result in mass acceptance of evolution. Yet getting the word out is proving disturbingly difficult. So the chatty pet theories about why creationism is a problem and what to do about it continue to dominate the national conversation and pro-creationist opinion remains rock steady (although the numbers of those who favor evolution without a god is rising along with the general increase of nonbelievers).

It's not just evolution. A classic example of conversational thinking by a scientist causing trouble was Linus Pauling's obsession with vitamin C. Many ordinary citizens are skeptical of scientists in general. When researchers offer up poorly sustained opinions on matters outside their firm knowledge base, it does not help the general situation.

So what can be done? In principle, the solution is simple enough. Scientists should be scientists. We should know better than to cough up committed but dubious opinion on subjects outside our expertise. This does not mean that scientists must limit their observations solely to their official field of research. Say a scientist is also a self-taught authority on baseball. By all means, ardently discuss that subject, the way Stephen Jay Gould used to.

I have long had an intense interest in the myths of World War II and can offer an excellent discourse on why the atom-bombing of Hiroshima and Nagasaki had pretty much nothing to do with ending the war, in case you're interested. (It was the Soviet attack on Japan that forced Hirohito to surrender to save his war-criminal's neck and keep Japan from being split into occupation zones.) But if they're asked about something they don't know a lot about, they should either decline to opine or qualify their observations by stating that the opinion is tentative and nonexpert.

In practical terms, the problem is, of course, that scientists are human beings like everyone else. So I'm not holding my breath waiting for us to achieve a level of factual discourse that will spread enlightenment to the masses. It's too bad, but very human. I have tried to cut down on throwing out idle commentary without qualifying its questionable reality, while being ardent about my statements only when I know I can back them up. Methinks I am fairly successful in this endeavor, and it does seem to keep me out of trouble.

BRICOLEUR

JAMES CROAK
Artist

French for "handyman" or "do-it-yourselfer," this word has migrated into art and philosophy recently, and savants would do well tossing it into their cognitive toolbox. A *bricoleur* is a talented tinkerer, the sort who can build anything out of anything: whack off a left-over drain pipe, fasten a loop of tin roofing, dab on some paint, and presto, a mailbox. If one peers closely, all the parts are still there—still a piece of roofing, a piece of pipe—but now the assembly exceeds the sum of the parts and is useful in a different way. In letters, a *bricoleur* is viewed as an intellectual MacGyver, tacking bits of his heritage to subcultures about him for a new meaning-producing pastiche.

Bricolage is not a new thing, but it has become a new way of understanding old things: epistemology, the Counter-Enlightenment, the endless parade of "isms" of the nineteenth and twentieth centuries. Marxism, Modernism, Socialism, Surrealism, Abstract Expressionism, Minimalism—the list is endless and often exclusive, each insisting that the other cannot be. The exegesis of these grand theories by deconstruction—substituting trace for presence—and similar activities during the past century shows these worldviews not as discoveries but as assemblies by creative *bricoleurs* working in the background, stapling together meaning-producing scenarios from textual bric-a-brac lying about.

Currently, encompassing worldviews in philosophy have been shelved, and master art movements of style and conclusion folded alongside them; no more isms are being run up the flagpole, because no one is saluting. Pluralism and modest descriptions of the world have become the activity of fine arts and letters, personalization and private worlds the Zeitgeist. The common prediction was that the loss of grand narrative would result in a descent into end-of-history purposelessness; instead, everywhere, the *bricoleurs* are busy manufacturing meaning-eliciting metaphor.

Motion Graphics, Bio-Art, Information Art, Net Art, Systems Art, Glitch Art, Hacktivism, Robotic Art, Relational Esthetics, and others—all current art movements tossed up by contemporary *bricoleurs* in an endless salad. Revisit nineteenth-century Hudson River landscape painting? Why not. Neo-Rodin, Post–New Media? A Mormon dabbling with the Frankfurt School? Next month. With the quest for universal validity suspended, there is a pronounced freedom to assemble lives filled with meaning from the nearby and at-hand. One just needs a *bricoleur*.

SCIENCE'S METHODS AREN'T JUST FOR SCIENCE

MARK HENDERSON
Science editor, The Times; *author*, 50 Genetics Ideas You Really Need to Know

Most people tend to think of science in one of two ways. It is a body of knowledge and understanding about the world: gravity, photosynthesis, evolution. Or it is the technology that has emerged from the fruits of that knowledge: vaccines, computers, cars. Science is both of these things, yet as Carl Sagan so memorably explained in *The Demon-Haunted World*, it is something else besides. It is a way of thinking, the best approach yet devised (if still an imperfect one) for discovering progressively better approximations of how things really are.

Science is provisional, always open to revision in light of new evidence. It is antiauthoritarian: Anybody can contribute, and anybody can be wrong. It seeks actively to test its propositions. And it is comfortable with uncertainty. These qualities give the scientific method unparalleled strength as a way of finding things out. Its power, however, is too often confined to an intellectual ghetto—those disciplines that have historically been considered "scientific."

Science as a method has great things to contribute to all sorts of pursuits beyond the laboratory. Yet it remains missing in action from far too much of public life. Politicians and civil servants too seldom appreciate how tools drawn from both the natural and social sciences can be used to design more effective policies, and even to win votes.

In education and criminal justice, for example, interventions are regularly undertaken without being subjected to proper evaluation. Both fields can be perfectly amenable to one of science's most potent techniques—the randomized controlled trial—yet these are seldom required before new initiatives are put into place. Pilots are often derisory in nature, failing even to collect useful evidence that could be used to evaluate a policy's success.

Sheila Bird, of the Medical Research Council, for instance, has criticized the UK's introduction of a new community sentence called the Drug Treatment and Testing Order, following pilots designed so poorly as to be worthless. They included too few subjects, they were not randomized, they did not properly compare the orders with alternatives, and judges were not even asked to record how they would otherwise have sentenced offenders.

The culture of public service could also learn from the self-critical culture of science. As Jonathan Shepherd, of the University of Cardiff, has pointed out, policing, social care, and education lack the cadre of practitioner-academics that has served medicine so well. There are those who do, and there are those who research; too rarely are they the same people. Police officers, teachers, and social workers are simply not encouraged to examine their own methods in the same way as doctors, engineers, and bench scientists. How many police stations run the equivalent of a journal club?

The scientific method and the approach to critical thinking it promotes are too useful to be kept back for "science" alone. If science can help us to understand the first microseconds of creation and the structure of the ribosome, it can surely improve understanding of how best to tackle the pressing social questions of our time.

THE GAME OF LIFE—AND LOOKING FOR GENERATORS

NICK BOSTROM
Director of the Future of Humanity Institute; professor, Faculty of Philosophy, University of Oxford

The Game of Life is a cellular automaton invented by the British mathematician John Horton Conway in 1970. Many will already be acquainted with Conway's invention. For those who aren't, the best way to familiarize yourself with it is to experiment with one of the many free implementations found on the Internet (or better yet, if you have at least rudimentary programming skills, make one yourself).

Basically, there is a grid, and each cell of the grid can be in either of two states: dead or alive. You start by seeding the grid with some initial distribution of live cells. Then you let the system evolve according to three simple rules.

Why is this interesting? Certainly, the Game of Life is not biologically realistic. It doesn't do anything useful. It isn't even really a game, in the ordinary sense of the word. But it's a brilliant demonstration platform for several important concepts—a virtual "philosophy of science laboratory." (The philosopher Daniel Dennett has expressed the view that it should be incumbent on every philosophy student to be acquainted with it.) It gives us a microcosm simple enough that we can easily understand how things are happening, yet with sufficient generative power to produce interesting phenomena.

By playing with the Game of Life for an hour, you can develop an intuitive understanding of the following concepts and ideas:

- *Emergent complexity*—How complex patterns can arise from very simple rules.

- *Basic dynamics concepts*—Such as the distinction between laws of nature and initial conditions.
- *Levels of explanation*—You quickly notice patterns (such as "gliders," which are a specific kind of pattern that crawls across the screen) arising that can be efficiently described in higher-level terms but are cumbersome to describe in the language of the basic physics (i.e., in terms of individual pixels being alive or dead) upon which the patterns supervene.
- *Supervenience*—This leads one to think about the relation between different sciences in the real world. Does chemistry, likewise, supervene on physics? Biology on chemistry? The mind on the brain?
- *Concept formation, and carving nature at its joints*—How and why we recognize certain types of patterns and give them names. For instance, in the Game of Life, you can distinguish "still lives," small patterns that are stable and unchanging; "oscillators," patterns that perpetually cycle through a fixed sequence of states; "spaceships," patterns that move across the grid (such as gliders); "guns," stationary patterns that send out an incessant stream of spaceships; and "puffer trains," patterns that move across the grid leaving debris behind. As you begin to form these and other concepts, the chaos on the screen gradually becomes more comprehensible. Developing concepts that carve nature at its joints is the first crucial step toward understanding, not only in the Game of Life but in science and in ordinary life as well.

At a more advanced level, we discover that the Game of Life is Turing complete. That is, it's possible to build a pattern that acts like a Universal Turing Machine (a computer that can simulate any other computer). Thus, any computable function could be implemented in the Game of Life—including perhaps a function that describes a universe like the one we inhabit. It's also possible to build a universal constructor in the Game of Life, a pattern that can build many types of complex objects, including copies of itself. Nonetheless, the structures that evolve into a Game of Life are different from those we find in the real world: Game of Life structures tend to be fragile, in the sense that changing a single cell will often cause them to dissolve. It is interesting to try to figure out exactly what it is about the rules of the Game of Life and the laws of physics that govern our own universe that accounts for these differences.

Conway's Game of Life is perhaps best viewed not as a single shorthand abstraction but rather as a generator of such abstractions. We get a whole bunch of useful abstractions—or at least a recipe for how to generate them—all for the price of one. And this points us to one especially useful shorthand abstraction: the strategy of Looking for Generators. We confront many problems. We can try to solve them one by one. But alternatively, we can try to create a generator that produces solutions to multiple problems.

Consider, for example, the challenge of advancing scientific understanding. We might make progress by directly tackling some random scientific problem. But perhaps we can make more progress by Looking for Generators and focusing our efforts on certain subsets of scientific problems—namely, those whose solutions would do most to facilitate the discovery of many other solutions. In this approach, we would pay most attention to innovations in methodology that can be widely applied; and to the development of scientific instruments that can enable many new experiments; and to improvements in institutional processes, such as peer review, that can help us make decisions about whom to hire, fund, or promote—decisions more closely reflecting true merit.

In the same vein, we would be extremely interested in developing effective biomedical cognitive enhancers and other ways of improving the human thinker—the brain being, after all, the generator par excellence.

ANECDOTALISM

ROBERT SAPOLSKY

Neuroscientist, Stanford University; author, Monkeyluv: And Other Essays on Our Lives as Animals

Various concepts come to mind for inclusion in that cognitive toolkit. "Emergence." Or, related to that, "The failure of reductionism": Mistrust the idea that if you want to understand a complex phenomenon, the only tool of science to use is to break it into its component parts, study them individually in isolation, and then glue the itty-bitty pieces back together. This doesn't often work, and increasingly it doesn't work for the most interesting and important phenomena out there. To wit: You have a watch that doesn't run correctly; often, you can fix it by breaking it down to its component parts and finding the gear that has had a tooth break (actually, I haven't a clue if there is any clock on Earth that still works this way). But if you have a cloud that doesn't rain, you don't break it down to its component parts. Ditto for a person whose mind doesn't work right. Or for going about understanding the problems of a society or ecosystem.

Related to that are terms like "synergy" and "interdisciplinary," but heaven save us from having to hear more about either of those words. There are now whole areas of science where you can't get a faculty position unless you work one of those words into the title of your job talk and have it tattooed on the small of your back.

Another useful scientific concept is "genetic vulnerability." One hopes this will find its way into everyone's cognitive toolkit, because its evil cousins "genetic inevitability" and "genetic determinism" are already deeply entrenched there, and with long long legacies of bad consequences. Everyone should be taught about work like that of Avshalom Caspi and colleagues, who looked at genetic polymorphisms related to various neurotransmitter systems associated with psychiatric disorders and antisocial behaviors. "Aha," far too many people will say, drawing on that nearly useless, misshapen tool of genetic determinism. "Have one of those polymorphisms and you're hosed by inevitability." And instead, what those studies beautifully demonstrate is how these polymorphisms carry essentially zero increased risk of those disorders, unless you grow up in particularly malign environments. Genetic determinism, my *tuchus*.

But the scientific concept I've chosen is one that's useful simply because it isn't a scientific concept: "anecdotalism." Every good journalist knows its power—start an article with statistics about foreclosure rates, or feature a family victimized by some bank? No-brainer. Display maps showing the magnitudes of refugees flowing out of Darfur or the face of one starving orphan in a camp? Obvious choice. Galvanize the readership.

But anecdotalism is potentially a domain of distortion as well. Absorb the lessons of science and cut saturated fats from your diet, or cite your friend's spouse's uncle who eats nothing but pork rinds and is still pumping iron at age 110? Depend on one of the foundations of the twentieth century's extension of life span and vaccinate your child, or obsess over a *National Enquirer*–esque horror story of one vaccination disaster and don't immunize? I shudder at the current potential for another case of anecdotalism: I write four days after the Arizona shooting of Gabrielle Giffords and nineteen other people by Jared Loughner. As of this writing, experts such as the esteemed psychiatrist Fuller Torrey are guessing that Loughner is a paranoid schizophrenic. And if this is true, this anecdotalism will give new legs to the tragic misconception that the mentally ill are more dangerous than the rest of us.

So maybe when I argue for anecdotalism going into everyone's cognitive toolkit, I am really arguing for two things to be incorporated: (a) an appreciation of how distortive it can be; and (b) recognition, in a salute to the work of people like Amos Tversky and Daniel Kahneman, of its magnetic pull, its cognitive satisfaction. As social primates complete with a region of the cortex specialized for face recognition, we find that the individual face—whether literal or metaphorical—has a special power. But unappealing, unintuitive patterns of statistics and variation generally teach us much more.

YOU CAN SHOW THAT SOMETHING IS DEFINITELY DANGEROUS BUT NOT THAT IT'S DEFINITELY SAFE

TOM STANDAGE

Digital editor, The Economist; *author,* An Edible History of Humanity

A wider understanding of the fact that you can't prove a negative would, in my view, do a great deal to upgrade the public debate around science and technology.

As a journalist, I have lost count of the number of times people have demanded that a particular technology be "proved to do no harm." This is, of course, impossible, in just the same way that proving that there are no black swans is impossible. You can look for a black swan (harm) in various ways, but if you fail to find one, that doesn't mean none exist: Absence of evidence is not evidence of absence.

All you can do is look again for harm in a different way. If you still fail to find it after looking in all the ways you can possibly think of, the question is still open: "lack of evidence of harm" means both "safe as far as we can tell" and "we still can't be sure if it's safe or not."

Scientists are often accused of logic-chopping when they point this out. But it would be immensely helpful to public discourse if there were a wider understanding that you can show that something is definitely dangerous, but you cannot show that it is definitely safe.

ABSENCE AND EVIDENCE

CHRISTINE FINN
Archaeologist; journalist; author, Artifacts: An Archaeologist's Year in Silicon Valley

I first heard the words "absence of evidence is not evidence of absence" as a first-year archaeology undergraduate. I now know it was part of Carl Sagan's retort against evidence from ignorance, but at the time the nonascribed quote was part of the intellectual toolkit offered by my professor to help us make sense of the process of excavation.

Philosophically this is a challenging concept, but at an archaeological site all became clear in the painstaking tasks of digging, brushing, and troweling. The concept was useful to remind us, as we scrutinized what was there, to take note of the possibility of what was not there. What we were finding, observing, and lifting were the material remains, the artifacts that had survived, usually as a result of their material or the good fortune of their deposition. There were barely recordable traces of what was there—the charcoal layer of a prehistoric hearth, for example—and others recovered in the washing or the lab, but this was still tangible evidence. What the concept brought home to us was the invisible traces, the material that had gone from our reference point in time but that still had a bearing in the context.

It was powerful stuff that stirred my imagination. I looked for more examples outside philosophy. I learned about the great Near Eastern archaeologist Leonard Woolley, who, when excavating the third millennium B.C. Mesopotamian palace at Ur, in modern-day Iraq, conjured up musical instruments from their absence. The evidence was the holes left in the excavation layers, the ghosts of wooden objects that had long since disappeared into time. He used this absence to advantage by making casts of the holes and realizing the instruments as reproductions. It struck me at the time that he was creating works of art. The absent lyres were installations he rendered as interventions and transformed into artifacts. More recently, the British artist Rachel Whiteread has made her name through an understanding of the absent form, from the cast of a house to the undersides and spaces of domestic interiors.

Recognizing the evidence of absence is not about forcing a shape on the intangible but about acknowledging a potency in the not-thereness. Taking the absence concept to be a positive idea, I suggest interesting things happen. For years, Middle Eastern archaeologists puzzled over the numerous isolated bathhouses and other structures in the deserts of North Africa. Where was the evidence of habitation? The clue was in the absence: The buildings were used by nomads, who left only camel tracks in the sand. Their habitations were ephemeral—tents that, if not taken away with them, were of such material that they would disappear into the sand. Observed again in this light, the aerial photos of desert ruins are hauntingly repopulated.

The absent evidence of ourselves is all around us, beyond the range of digital traces.

When my parents died and I inherited their house, the task of clearing their rooms was both emotional and archaeological. The last mantelpiece in the sitting room had accreted over thirty-five years of married life a midden of photos, ephemera, beachcombing trove, and containers of odd buttons and old coins. I wondered what a stranger—maybe a forensic scientist or traditional archaeologist—would make of this array if the narrative had been woven simply from the tangible evidence. But as I took the assemblage apart in a charged moment, I felt there was a whole lot of no-thing that was coming away with it. Something invisible and unquantifiable, which had been holding these objects in that context.

I recognized the feeling and cast my memory back to my first archaeological excavation. It was of a long-limbed hound, one of those "fine hunting dogs" that the classical writer Strabo described as being traded from ancient Britain into the Roman world. As I knelt in the two-thousand-year-old grave, carefully removing each tiny bone, as if engaged in a sculptural process, I felt the presence of something absent. I could not quantify it, but it was that unseen "evidence" that, it seemed, had given the dog its dogness.

PATH DEPENDENCE

JOHN McWHORTER
Linguist; cultural commentator; senior fellow, Manhattan Institute; lecturer, Department of English & Comparative Literature, Columbia University; author, What Language Is (And What It Isn't and What It Could Be)

In an ideal world, all people would spontaneously understand that what political scientists call *path dependence* explains much more of how the world works than is apparent. "Path dependence" refers to the fact that often something that seems normal or inevitable today began with a choice that made sense at a particular time in the past but survived despite the eclipse of its justification, because, once it had been established, external factors discouraged going into reverse to try other alternatives.

The paradigm example is the seemingly illogical arrangement of letters on typewriter keyboards. Why not just have the letters in alphabetical order, or arrange them so that the most frequently occurring ones are under the strongest fingers? In fact, the first typewriter tended to jam when typed on too quickly, so its inventor deliberately concocted an arrangement that put "A" under the ungainly little finger. In addition, the first row was provided with all of the letters in the word "typewriter," so that salesmen, new to typing, could type the word using just one row.

Quickly, however, mechanical improvements made faster typing possible, and new keyboards placing letters according to frequency were presented. But it was too late: There was no going back. By the 1890s, typists across America were used to QWERTY keyboards, having learned to zip away on new versions of them that did not stick so easily. Retraining them would have been expensive and, ultimately, unnecessary, so QWERTY was passed down the generations, and even today we use the queer QWERTY configuration on computer keyboards, where jamming is a mechanical impossibility.

The basic concept is simple, but in general estimation tends to be processed as the province of "cute" stories like the QWERTY one, rather than explaining a massive weight of scientific and historical processes. Instead, the natural tendency is to seek explanations for modern phenomena in modern conditions.

One may assume that cats cover their waste out of fastidiousness, when the same creature will happily consume its own vomit and then jump on your lap. Cats do the burying as an instinct from their wild days, when the burial helped avoid attracting predators, and there is no reason for them to evolve out of the trait now (to pet owners' relief). I have often wished there were a spontaneous impulse among more people to assume that path-dependence–style explanations are as likely as jerry-rigged present-oriented ones. For one thing, that the present is based on a dynamic mixture of extant and ancient conditions is simply more interesting than assuming that the present is (mostly) all there is, with history as merely "the past," interesting only for seeing whether something that happened then could now happen again—which is different from path dependence.

For example, path dependence explains a great deal about language which is otherwise attributed to assorted just-so explanations. Much of the public embrace of the idea that one's language channels how one thinks is based on this kind of thing. Robert McCrum celebrates English as "efficient" in its paucity of suffixes of the kind that complexify most European languages. The idea is that this is rooted in something in its speakers' spirit that would have propelled them to lead the world via exploration and the Industrial Revolution.

But English lost its suffixes starting in the eighth century A.D., when Vikings invaded Britain and so many of them learned the language incompletely that children started speaking it that way. After that, you can't create gender and conjugation out of thin air—there's no going back until gradual morphing re-creates such things over eons of time. That is, English's current streamlined syntax has nothing to do with any present-day condition of the spirit nor with any even four centuries ago. The culprit is path dependence, as are most things about how a language is structured.

Or we hear much lately about a crisis in general writing skills, supposedly due to e-mail and texting. But there is a circularity here: Why, precisely, could people not write e-mails and texts with the same "writerly" style that people once couched letters in? Or we hear of a vaguely defined effect of television, although kids were curled up endlessly in front of the tube starting in the fifties, long before the eighties when outcries of this kind first took on their current level of alarm, in the National Commission on Excellence in Education's report *A Nation at Risk*.

Once again, the presentist explanation does not cohere, whereas one based on an earlier historical development that there is no turning back from does. Public American English began a rapid shift from cosseted to less formal "spoken" style in the sixties, in the wake of cultural changes amid the counterculture. This sentiment directly affected how language arts textbooks were composed, the extent to which any young person was exposed to an old-fashioned formal "speech," and attitudes toward the English language heritage in general. The result: a linguistic culture stressing the terse, demotic, and spontaneous. After just one generation minted in this context, there was no way to go back. Anyone who decided to communicate in the grandiloquent phraseology of yore would sound absurd and be denied influence or exposure. Path dependence, then, identifies this cultural shift as the cause of what dismays, delights, or just interests us in how English is currently used—and reveals television, e-mail and other technologies as merely epiphenomenal.

Most of life looks path-dependent to me. If I could create a national educational curriculum from scratch, I would include the concept as one taught to young people as early as possible.

INTERBEING

SCOTT D. SAMPSON
Dinosaur paleontologist, evolutionary biologist, science communicator; author, Dinosaur Odyssey: Fossil Threads in the Web of Life

Humanity's cognitive toolkit would greatly benefit from adoption of "interbeing," a concept that comes from the Vietnamese Buddhist monk Thich Nhat Hanh. In his words:

> If you are a poet, you will see clearly that there is a cloud floating in [a] sheet of paper. Without a cloud, there will be no rain; without rain, the trees cannot grow; and without trees, we cannot make paper. The cloud is essential for the paper to exist. If the cloud is not here, the sheet of paper cannot be here either. . . . "Interbeing" is a word that is not in the dictionary yet, but if we combine the prefix "inter-" with the verb "to be," we have a new verb, inter-be. Without a cloud, we cannot have a paper, so we can say that the cloud and the sheet of paper *inter-are*. . . . To be is to inter-be. You cannot just be by yourself alone. You have to inter-be with every other thing. This sheet of paper is, because everything else is.

Depending on your perspective, the above passage may sound like profound wisdom or New Age mumbo-jumbo. I would like to propose that interbeing is a robust scientific fact—at least, insofar as such things exist—and, further, that this concept is exceptionally critical and timely.

Arguably the most cherished and deeply ingrained notion in the Western mind-set is the separateness of our skin-encapsulated selves—the belief that we can be likened to isolated, static machines. Having externalized the world beyond our bodies, we are consumed by thoughts of furthering our own ends and protecting ourselves. Yet this deeply rooted notion of isolation is illusory, as evidenced by our constant exchange of matter and energy with the "outside" world. At what point did your last breath of air, sip of water, or bite of food cease to be part of the outside world and become you? Precisely when did your exhalations and wastes cease being you? Our skin is as much permeable membrane as barrier—so much so that, like a whirlpool, it is difficult to discern where "you" end and the remainder of the world begins. Energized by sunlight, life converts inanimate rock into nutrients, which then pass through plants, herbivores, and carnivores before being decomposed and returned to the inanimate Earth, beginning the cycle anew. Our internal metabolisms are intimately interwoven with this Earthly metabolism; one result is the replacement of every atom in our bodies every seven years or so.

You might counter with something like, "OK, sure, everything changes over time. So what? *At any given moment*, you can still readily separate self from other."

Not quite. It turns out that "you" are not one life-form—that is, one self—but many. Your mouth alone contains more than seven hundred distinct kinds of bacteria. Your skin and eyelashes are equally laden with microbes, and your gut houses a similar bevy of bacterial sidekicks. Although this still leaves several bacteria-free regions in a healthy body—for example, brain, spinal cord, and bloodstream—current estimates indicate that your physical self possesses about 10 trillion human cells and about 100 trillion bacterial cells. In other words, *at any given moment*, your body is about 90 percent nonhuman, home to many more life-forms than the number of people presently living on Earth; more even than the number of stars in the Milky Way galaxy! To make things more interesting still, microbiological research demonstrates that we are utterly dependent on this ever-changing bacterial parade for all kinds of "services," from keeping intruders at bay to converting food into usable nutrients.

So, if we continually exchange matter with the outside world, if our bodies are completely renewed every few years, and if each of us is a walking colony of trillions of largely symbiotic life-forms, exactly what is this self that we view as separate? You are not an isolated being. Metaphorically, to follow current bias and think of your body as a machine is not only inaccurate but destructive. Each of us is far more akin to a whirlpool, a brief, ever-shifting concentration of energy in a vast river that has been flowing for billions of years. The dividing line between self and other is, in many respects, arbitrary; the "cut" can be made at many places, depending on the metaphor of self that one adopts. We must learn to see ourselves not as isolated but as permeable and interwoven—selves within larger selves, including the species self (humanity) and the biospheric self (life). The interbeing perspective encourages us to view other life-forms not as objects but subjects, fellow travelers in the current of this ancient river. On a still more profound level, it enables us to envision ourselves and other organisms not as static "things" at all but as *processes* deeply and inextricably embedded in the background flow.

One of the greatest obstacles confronting science education is the fact that the bulk of the universe exists either at extremely large scales (e.g., planets, stars, and galaxies) or extremely small scales (e.g., atoms, genes, cells) well beyond the comprehension of our (unaided) senses. We evolved to sense only the middle ground, or "mesoworld," of animals, plants, and landscapes. Yet, just as we have learned to accept the nonintuitive, scientific insight that the Earth is not the center of the universe, so too must we now embrace the fact that we are not outside or above nature but fully enmeshed within it. Interbeing, an expression of ancient wisdom backed by science, can help us comprehend this radical ecology, fostering a much-needed transformation in mind-set.

THE OTHER

DIMITAR SASSELOV
Professor of astronomy; director, Harvard Origins of Life Initiative

The concept of "otherness" or "the Other" is part of how a human being perceives his or her own identity: "How do I relate to others?" is a part of what defines the self and is constituent in self-consciousness. It is a philosophical concept widely used in psychology and social science. Recent advances in the life and physical sciences have made possible new and even unexpected expansions of this concept. The map of the human genome and of the diploid genomes of individuals; the map of our geographic spread; the map of the Neanderthal genome—these are new tools to address the age-old issues of human unity and diversity. Reading the life code of DNA does not stop there; it places humans in the vast and colorful mosaic of earthly life. "Otherness" is seen in a new light. Our microbiomes—the trillions of microbes on and in each of us, and essential to our physiology, become part of our selves.

Astronomy and space science are intensifying the search for life on other planets—from Mars and the outer reaches of the solar system to Earth-like planets and super-Earths orbiting other stars. The chances of success may hinge on our understanding of the possible diversity of the chemical basis of life itself: "otherness" not among DNA-encoded species but among life-forms using different molecules to encode traits. Our 4-billion–year-old heritage of molecular innovation and design versus "theirs." This is a cosmic encounter that we might experience first in our laboratories. Last year's creation of JCVI-syn1.0—the first bacterial cell controlled completely by a synthetic genome—is a prelude to this brave new field.

It is probably timely to ponder "otherness" and its wider meaning, as we embark on a new age of exploration. As T. S. Eliot predicted in "Little Gidding," we might arrive where we started and know our *self* for the first time.

ECOLOGY

BRIAN ENO
Artist; composer; recording producer: U2, Coldplay, Talking Heads, Paul Simon; recording artist; author, A Year with Swollen Appendices: Brian Eno's Diary

That idea, or bundle of ideas, seems to me the most important revolution in general thinking in the last hundred and fifty years. It has given us a whole new sense of who we are, where we fit, and how things work. It has made commonplace and intuitive a type of perception that used to be the province of mystics—the sense of wholeness and interconnectedness.

Beginning with Copernicus, our picture of a semidivine humankind perfectly located at the center of the universe began to falter: We discovered that we live on a small planet circling a medium-sized star at the edge of an average galaxy. And then, following Darwin, we stopped being able to locate ourselves at the center of life. Darwin gave us a matrix upon which we could locate life in all its forms, and the shocking news was that we weren't at the center of that, either—just another species in the innumerable panoply of species, inseparably woven into the whole fabric (and not an indispensable part of it either). We have been cut down to size, but at the same time we have discovered ourselves to be part of the most unimaginably vast and beautiful drama called Life.

Before "ecology" we understood the world in the metaphor of a pyramid—a hierarchy, with God at the top, Man a close second, and, sharply separated, a vast mass of life and matter beneath. In that model, information and intelligence flowed in one direction only—from the intelligent top to the "base" bottom—and, as masters of the universe, we felt no misgivings about exploiting the lower reaches of the pyramid.

The ecological vision has changed that: We now increasingly view life as a profoundly complex weblike system with information running in all directions, and instead of a single hierarchy we see an infinity of nested and codependent hierarchies—and the complexity of all this is, in and of itself, creative. We no longer need the idea of a superior intelligence outside the system; the dense field of intersecting intelligences is fertile enough to account for all the incredible beauty of "creation."

The ecological view isn't confined to the organic world. Along with it comes a new understanding of how intelligence itself comes into being. The classical picture saw Great Men with Great Ideas . . . but now we tend to think more in terms of fertile circumstances, wherein uncountable numbers of minds contribute to a river of innovation. It doesn't mean we cease to admire the most conspicuous of these—but that we understand them as effects as much as causes. This has ramifications for the way we think about societal design, about crime and conflict, education, culture, and science.

That in turn leads to a reevaluation of the various actors in the human drama. When we realize that the cleaners and the bus drivers and the primary-school teachers are as much a part of the story as the professors and the celebrities, we will start to accord them the respect they deserve.

DUALITIES

STEPHON H. ALEXANDER
Associate professor of physics, Haverford College

In the northeast Bronx, I walk through a neighborhood I once feared going into, this time with a big smile on my face. This is because I can quell the bullies with a new slang word in our dictionary, "dual." As I approach the 2 train stop on East 225th Street, the bullies await me. I say, "Yo, what's the dual?" The bullies embrace me with a pound followed by a high five. I make my train.

In physics, one of the most beautiful yet underappreciated ideas is that of duality. A duality allows us to describe a physical phenomenon from two different perspectives; often a flash of creative insight is needed to find both. However, the power of the duality goes beyond the apparent redundancy of description. After all, why do I need more than one way to describe the same thing? There are examples in physics where either description of a phenomenon fails to capture its entirety. Properties of the system "beyond" the individual descriptions "emerge." I will provide two beautiful examples of how dualities manage to yield emergent properties, and I will end with a speculation.

Most of us know about the famous wave-particle duality in quantum mechanics, which allows the photon (and the electron) to attain their magical properties to explain all of the wonders of atomic physics and chemical bonding. The duality states that matter (such as the electron) has both wavelike and particle-like properties, depending on the context. What's weird is *how* quantum mechanics manifests the wave-particle duality. According to the traditional Copenhagen interpretation, the wave is a traveling oscillation of possibility that the electron can be realized somewhere as a particle.

Life gets strange in the example of quantum tunneling, where the electron can penetrate a barrier only because of the electron's wavelike property. Classical physics tells us that an object will not surmount a barrier (like a hill) if its total kinetic energy is less than the potential energy of the barrier. However, quantum mechanics predicts that a particle can penetrate (or tunnel) through a barrier even when the particle's kinetic energy is less than the potential energy of the barrier. This effect occurs every time you use a flash drive or a CD player.

Most people assume that the conduction of electrons in a metal is a well-understood property of classical physics. But when we look deeper, we realize that conduction happens because of the wavelike nature of the electrons. We call the electron waves moving through the periodic lattice of a metal Bloch waves. When the electrons' Bloch waves constructively interfere, we get conduction. Moreover, the wave-particle duality takes us further, predicting superconductivity: How it is that electrons (and other spin-half particles, like quarks) can conduct without resistance?

Nowadays, in my field of quantum gravity and relativistic cosmology, theorists are exploiting another type of duality to address unresolved questions. This holographic duality was pioneered by Leonard Susskind and Gerard 't Hooft, and later it found a home in the form of the AdS/CFT (anti-de-Sitter space/conformal field theory) duality conceived by Juan Maldacena. This posits that the phenomenon of quantum gravity is described, on the one hand, by an ordinary gravitational theory (a beefed-up version of Einstein's general relativity). On the other hand, a dual description of quantum gravity is described by a nongravitational physics with a space-time of one lower dimension. We are left to wonder, in the spirit of the wave-particle duality, what new physics we will glean from this type of duality.

The holographic duality seems to persist in other approaches to quantum gravity, such as loop quantum gravity, and researchers are still exploring the true meaning behind holography and the potential predictions for experiments.

Dualities allow us to understand and make use of properties in physics that go beyond a singular lens of analysis. Might we wonder if duality can transcend its role in physics and enter other fields? The dual of time will tell.

DUALITIES

AMANDA GEFTER

Books & Arts editor, New Scientist; *founder and editor*, CultureLab

It is one of the stranger ideas to emerge from recent physics. Take two theories that describe utterly dissimilar worlds—worlds with different numbers of dimensions, different geometries of space-time, different building blocks of matter. Twenty years ago, we'd say those are indisputably disparate and mutually exclusive worlds. Today, there's another option: Two radically different theories might be *dual* to one another—that is, they might be two very different manifestations of the same underlying reality.

Dualities are as counterintuitive a notion as they come, but physics is riddled with them. When physicists looking to unite quantum theory with gravity found themselves with five very different but equally plausible string theories, it was an embarrassment of riches—everyone was hoping for one "theory of everything," not five. But duality proved to be the key ingredient. Remarkably, all five string theories turned out to be dual to one another, different expressions of a single underlying theory.

Perhaps the most radical incarnation of duality was discovered in 1997 by the theoretical physicist Juan Maldacena. Maldacena found that a version of string theory in a bizarrely shaped universe with five large dimensions is mathematically dual to an ordinary quantum theory of particles living on that universe's four-dimensional boundary. Previously, one could argue that the world was made up of particles *or* that the world was made up of strings. Duality transformed *or* into *and*—mutually exclusive hypotheses, both equally true.

In everyday language, "duality" means something else. It connotes a stark dichotomy: male and female, east and west, light and darkness. Embracing the physicist's meaning of duality, however, can provide us with a powerful new metaphor, a one-stop shorthand for the idea that two very different things might be equally true. As our cultural discourse is becoming increasingly polarized, the notion of duality is both more foreign and more necessary than ever. In our daily cognitive toolkit, it could serve as a potent antidote to our typically Boolean, two-valued, zero-sum thinking—where statements are either true or false, answers are yes or no, and if I'm right, then you're wrong. With duality, there's a third option. Perhaps my argument is right and yours is wrong; perhaps your argument is right and mine is wrong; or, just maybe, our opposing arguments are dual to one another.

That's not to say that we ought to descend into some kind of relativism, or that there are no singular truths. It is to say, though, that truth is far more subtle than we once believed, and that it shows up in many guises. It is up to us to recognize it in all its varied forms.

THE PARADOX

ANTHONY AGUIRRE
Associate professor of physics, University of California–Santa Cruz

Paradoxes arise when one or more convincing truths contradict each other, clash with other convincing truths, or violate unshakable intuitions. They are frustrating, yet beguiling. Many see virtue in avoiding, glossing over, or dismissing them. Instead we should seek them out. If we find one, sharpen it, push it to the extreme, and hope that the resolution will reveal itself, for with that resolution will invariably come a dose of Truth.

History is replete with examples, and with failed opportunities. One of my favorites is Olber's paradox. Suppose the universe were filled with an eternal, roughly uniform distribution of shining stars. Faraway stars would look dim, because they take up a tiny angle on the sky, but within that angle they are as bright as the sun's surface. Yet in an eternal and infinite (or finite but unbounded) space, every direction would lie within the angle taken up by some star. The sky would be alight like the surface of the sun. Thus, a simple glance at the dark night sky reveals that the universe must be dynamic: expanding, or evolving. Astronomers grappled with this paradox for several centuries, devising unworkable schemes for its resolution. Despite at least one correct view (by Edgar Allan Poe!), the implications never really permeated even the small community of people thinking about the fundamental structure of the universe. And so it was that Albert Einstein, when he went to apply his new theory to the universe, sought an eternal and static model that could never make sense, introduced a term into his equations which he later called his greatest blunder, and failed to invent the Big Bang theory of cosmology.

Nature appears to contradict itself with the utmost rarity, and so a paradox can be an opportunity for us to lay bare our cherished assumptions and discover which of them we must let go. But a good paradox can take us further, to reveal that not just the assumptions but the very modes of thinking we used in creating the paradox must be replaced. Particles and waves? Not truth, just convenient models. The same number of integers as perfect squares of integers? Not crazy, though you might be, if you invent cardinality. "This sentence is false." And so, says Gödel, might be the foundations of any formal system that can refer to itself. The list goes on.

What next? I've got a few big ones I'm wrestling with. How can the second law of thermodynamics arise, unless cosmological initial conditions are fine-tuned in a way we would never accept in any other theory or explanation of anything? How do we do science, if the universe is infinite and every outcome of every experiment occurs infinitely many times?

What impossibility is nagging at you?

HUNTING FOR ROOT CAUSE: THE HUMAN "BLACK BOX"

ERIC TOPOL

Professor of translational genomics, Scripps Research Institute; cardiologist, Scripps Clinic

Root-cause analysis is an attractive concept for certain matters in industry, engineering, and quality control. A classic application is to determine why a plane crashed by finding the "black box"—the tamper-proof event-data recorder. Even though this box is usually bright orange, the term symbolizes the sense of dark matter, a container with critical information to help illuminate what happened. Getting the black-box audio recording is just one component of a root-cause analysis of why a plane goes down.

Each of us is gradually being morphed into an event-data recorder by virtue of our digital identity and presence on the Web. Not only do we post our own data, sometimes unwittingly, but also others post information about us, and all of this is permanently archived. In that way, it is close to tamper-proof. With increasing use of biosensors, high-resolution imaging (just think of our current cameras and video recording, let alone our digital medical imaging), and DNA sequencing, the human event-data recorder will be progressively enriched.

In our busy, networked lives, with constant communication, streaming, and distraction, the general trend has moved away from acquiring deep knowledge of why something happened. This is best exemplified in health and medicine. Physicians rarely seek root causes. If a patient has a common condition, such as high blood pressure, diabetes, or asthma, he or she is put on a prescription drug without any attempt at ascertaining why the individual crashed—certainly a new chronic medical condition can be likened to such an event. There are usually specific reasons for these disorders, but they are not hunted down. Taken to an extreme, when an individual dies and the cause is not known, it is now exceedingly rare that an autopsy is performed. Doctors have generally caved in their quest to define root cause, and they are fairly representative of most of us. Ironically, this is happening at a time when there is unprecedented capability for finding the explanation. But we're just too busy.

So to tweak our cognitive performance in the digital world, where there is certainly no shortage of data, it's time to use it to understand, as fully as possible, why unexpected or unfavorable things happen. Or even why something great transpired. It's a prototypic scientific concept that has all too often been left untapped. Each person is emerging as an extraordinary event recorder and part of the Internet of all things. Let's go deep. Nothing unexplained these days should go without a hunt.

PERSONAL DATA MINING

DAVID ROWAN
Editor, Wired *magazine's UK edition*

From the dawn of civilization until 2003, former Google CEO Eric Schmidt is fond of saying, humankind generated five exabytes of data. Now we produce five exabytes every two days—and the pace is accelerating. In our post-privacy world of pervasive social-media sharing, GPS tracking, cellphone-tower triangulation, wireless sensor monitoring, browser-cookie targeting, face-recognition detecting, consumer-intention profiling, and endless other means by which our personal presence is logged in databases far beyond our reach, citizens are largely failing to benefit from the power of all this data to help them make smarter decisions. It's time to reclaim the concept of data mining from the marketing industry's microtargeting of consumers, the credit-card companies' antifraud profiling, the intrusive surveillance of state-sponsored Total Information Awareness. We need to think more about mining our own output to extract patterns that turn our raw personal data stream into predictive, actionable information. All of us would benefit if the idea of personal data mining were to enter popular discourse.

Microsoft saw the potential back in September 2006, when it filed United States Patent application number 20,080,082,393 for a system of "personal data mining." Having been fed personal data provided by users themselves or gathered by third parties, the technology would then analyze it to "enable identification of opportunities and/or provisioning of recommendations to increase user productivity and/or improve quality of life." You can decide for yourself whether you trust Redmond with your lifelog, but it's hard to fault the premise: The personal data mine, the patent states, would be a way "to identify relevant information that otherwise would likely remain undiscovered."

Both I as a citizen and society as a whole would gain if individuals' personal datastreams could be mined to extract patterns upon which we could act. Such mining would turn my raw data into predictive information that can anticipate my mood and improve my efficiency, make me healthier and more emotionally intuitive, reveal my scholastic weaknesses and my creative strengths. I want to find the hidden meanings, the unexpected correlations that reveal trends and risk factors of which I had been unaware. In an era of oversharing, we need to think more about data-driven self-discovery.

A small but fast-growing self-tracking movement is already showing the potential of such thinking, inspired by Kevin Kelly's quantified self and Gary Wolf's data-driven life. With its mobile sensors and apps and visualizations, this movement is tracking and measuring exercise, sleep, alertness, productivity, pharmaceutical responses, DNA, heartbeat, diet, financial expenditure—and then sharing and displaying its findings for greater collective understanding. It is using its tools for clustering, classifying, and discovering rules in raw data, but mostly it is simply quantifying that data to extract signals—information—from the noise.

The cumulative rewards of such thinking will be altruistic rather than narcissistic, whether in pooling personal data for greater scientific understanding (23andMe) or in propagating user-submitted data to motivate behavior change in others (traineo). Indeed, as the work of Daniel Kahneman, Daniel Gilbert, and Christakis and Fowler demonstrate so powerfully, accurate individual-level data tracking is key to understanding how human happiness can be quantified, how our social networks affect our behavior, how diseases spread through groups.

The data is already out there. We just need to encourage people to tap it, share it, and corral it into knowledge.

PARALLELISM IN ART AND COMMERCE

SATYAJIT DAS
Expert, financial derivatives and risk; author, Traders, Guns & Money: Knowns and Unknowns in the Dazzling World of Derivatives *and* Extreme Money

Confluence of factors is highly influential in setting off changes in complex systems. A common example is in risk—the "Swiss cheese" theory. Losses occur only if all controls fail and the holes in the Swiss cheese align.

Confluence—the coincidence of events in a single setting—is well understood. But parallel developments in different settings or disciplines can also be influential in shaping events. A coincidence of similar logic and processes in seemingly unrelated activities provides indications of likely future developments and risks. The ability to better recognize parallelism would improve our cognitive processes.

Economic forecasting is dismal, prompting John Kenneth Galbraith to remark that economists were put on Earth only to make astrologers look good. Few economists anticipated the current financial problems. However, the art market proved remarkably accurate in anticipating developments—especially the market in the work of Damien Hirst, the best known of a group of artists dubbed YBAs (Young British Artists).

Hirst's most iconic work—*The Physical Impossibility of Death in the Mind of Someone Living*—is a fourteen-foot tiger shark preserved in formaldehyde and immersed in a vitrine weighing over two tons. The advertising guru Charles Saatchi bought it for £50,000. In December 2004, Saatchi sold the work to Steven Cohen, founder and principal of the über hedge fund SAC Capital Advisors, which manages $16 billion. Cohen is thought to have paid $12 million for *The Physical Impossibility of Death in the Mind of Someone Living*, although there are allegations that it was "only" $8 million. In June 2007, Damien Hirst tried to sell a life-size platinum cast of a human skull encrusted with £15 million worth of 8,601 pavé-set industrial diamonds that weighed a total of 1,106 carats, including a 52.4-carat pink diamond in the center of the forehead valued at £4 million. *For the Love of God* was a memento mori (in Latin: *Remember you must die*). The work was offered for sale at £50 million, as part of Hirst's "Beyond Belief" show. In September 2007, *For the Love of God* was sold to Hirst and some investors for full price, for later resale.

The sale of *The Physical Impossibility of Death in the Mind of Someone Living* marked the final phase of the irresistible rise of markets. The failure of *For the Love of God* to sell marked its zenith as clearly as any economic marker.

Parallelism exposes common thought processes and similar valuation approaches to unrelated objects. Hirst was the artist of choice for conspicuously consuming hedge-fund managers, who were getting very rich. Inflated prices suggested the presence of irrational excess. The nature of sought-after Hirst pieces, and even their titles, provided an insight into the hubristic self-image of financiers. With its jaws gaping, poised to swallow its prey, *The Physical Impossibility of Death in the Mind of Someone Living* mirrored the killer instincts of hedge funds, feared predators in financial markets. Cohen is quoted as saying that he "liked the whole fear factor." The work of Japanese artist Takashi Murakami provides confirmation. Inspired by Hokusai's famous nineteenth-century woodblock print *The Great Wave off Kanagawa*, Murakami's 727 paintings showed Mr. DOB, a post-nuclear Mickey Mouse character, as a god riding on a cloud or a shark surfing on a wave. The first 727 is owned by New York's Museum of Modern Art, the second by Steven Cohen.

Parallelism is also evident in the causes underlying several crises facing humanity. It is generally acknowledged that high levels of debt were a major factor in the ongoing global financial crisis. What is missed is that the logic of debt is similar to one underlying other problematic issues. There is a striking similarity between the problems of the financial system, irreversible climate change, and shortages of vital resources like oil, food, and water. Economic growth and wealth were based on borrowed money. Debt allowed society to borrow from the future. It accelerated consumption, as debt is used to buy something today against the uncertain promise of paying back the borrowing in the future. Society polluted the planet, creating changes in the environment that are difficult to reverse. Underpriced, natural, finite resources were wantonly utilized, without proper concern about conservation.

In each area, society borrowed from, and pushed problems into, the future. Current growth and short-term profits were pursued at the expense of risks not immediately evident and which would emerge later.

To dismiss this as short-term thinking and greed is disingenuous. A crucial cognitive factor underlying the approach was a similar process of problem solving—borrowing from and pushing problems further into the future. This process was consistently applied across different problems, without consideration of its relevance, applicability, or desirability. Where such parallelism exists, it feeds on itself, potentially leading to total systemic collapse.

Recognition and understanding of parallelism is one way to improve our cognitive toolkit. It may provide a better mechanism for predicting specific trends. It may also enable people to increase dialectic richness, drawing on different disciplines. It requires overcoming highly segmented and narrow educational disciplines, rigid institutional structures, and restricted approaches to analysis and problem solving.

INNOVATION

LAURENCE C. SMITH

Professor of geography and earth & space sciences, University of California–Los Angeles; author, The World in 2050: Four Forces Shaping Civilization's Northern Future

As scientists, we're sympathetic to this year's *Edge* Question. We've asked it of ourselves before, many times, after fruitless days lost at the lab bench or the computer. If only our brains could find a new way to process the delivered information faster, to interpret it better, to align the world's noisy torrents of data in a crystalline moment of clarity. In a word, for our brains to forgo their familiar thought sequences and innovate.

To be sure, the word "innovate" has become something of a cliché. Tenacious CEOs, clever engineers, and restless artists come to mind, before the methodical, data-obsessed scientist. But how often do we consider the cognitive role of innovation in the supposedly bone-dry world of hypothesis testing, mathematical constraints, and data-dependent empiricism?

In the world of science, innovation stretches the mind to find an explanation when the universe wants to hold on to its secrets just a little longer. This can-do attitude is made all the more valuable, not less, in a world constrained by ultimate barriers—such as continuity of mass and energy, absolute zero, or the Clausius-Clapeyron relation. Innovation is a critical enabler of discovery around and outside of these bounds. It is the occasional architect of that rare, wonderful breakthrough made even when the tide of scientific opinion is against you.

A reexamination of this word from the scientific perspective reminds us of the extreme power of this cognitive tool, one that most people possess already. Through innovation, we all can transcend social, professional, political, scientific, and most important, personal limits. Perhaps we might all put it to more frequent use.

THE GIBBS LANDSCAPE

KEVIN HAND
Planetary scientist, Jet Propulsion Laboratory

Biology is rarely wasteful. Sure, on the individual organism level there is plenty of waste involved with reproduction and other activities (think of all the fruit on a tree or the millions of sperm that lose out in the race to the egg). But on the ecosystem level, one bug's trash is another bug's treasure—provided that some useful energy can still be extracted by reacting that trash with something else in the environment. The food chain is not a simple linear staircase of predator-prey relationships; it is a complex fabric of organisms large, small, and microscopic interacting with one another and with the environment to tap every possible energetic niche.

Geobiologists and astrobiologists can measure and map this energy—referred to as Gibbs free energy. Doing so is useful for assessing the energetic limits of life on Earth and for assessing potentially habitable regions on other worlds. In an ecosystem, Gibbs free energy—named for its discoverer, the late-nineteenth-century scientist J. Willard Gibbs—is the energy in a biochemical reaction that is available to do work. It's the energy left over after producing some requisite waste heat and a dollop or two of entropy. This energy to do work is harnessed by biological systems for activities like making repairs, growing, and reproducing. For a given metabolic pathway used by life—for example, reacting carbohydrates with oxygen—we can measure how many joules are available to do work per mole of reactants. Humans, and essentially all the animals you know and love, typically harness a couple thousand kilojoules per mole by burning food with oxygen.

Microbes have figured out all sorts of ways to harness Gibbs free energy by combining various gases, liquids, and rocks. Measurements by Tori Hoehler and colleagues at NASA Ames Research Center on methane-generating and sulfate-eating microbes indicate that the limit for life may be about 10 kilojoules per mole. Within a given environment, there may be many chemical pathways in operation, and if there is an open energetic niche, chances are that life will find a way to fill it. Biological ecosystems can be mapped as a landscape of reactions and pathways for harnessing energy. This is the Gibbs landscape.

Civilizations and the rise of industrial and technological ecosystems bring a new challenge to our understanding of the dynamic between energy needs and energy resources. The Gibbs landscape provides a shorthand abstraction for conceptualizing this dynamic. We can imagine any given city, country, or continent overlain with a map of energy available to do work. This includes, but extends beyond, the chemical-energy framework used in the context of biological ecosystems. For instance, automobiles with internal-combustion engines metabolize gasoline with air. Buildings eat the electricity supplied by power plants or rooftop solar panels. Every component in modern industrial society occupies some niche in the landscape.

But importantly, many of the Gibbs landscapes in place today are rife with unoccupied niches. The systems we have designed and built are inefficient and incomplete in the utilization of energy to do the work of civilization's ecosystems. Much of what we have designed excels at producing waste heat with little concern for optimizing work output. From lights that remain on all night to landfills that contain discarded resources, the Gibbs landscapes of today offer much room for technological innovation and evolution. The Gibbs landscape also provides a way to visualize untapped capacity for doing work—wind, solar, hydroelectric, tides, and geothermal are just a few of the layers. Taken together, these layers show us where and how we can work to close the loops and connect the dangling threads of our nascent technological civilization.

When you start to view the world around you with Gibbsian eyes, you see the untapped potential in many of our modern technological and industrial ecosystems. It's disturbing at first, because we've done such a poor job, but the marriage between civilization and technology is young. The landscape provides much reason for hope, as we continue to innovate and strive to reach the balance and continuity that has served complex biological ecosystems so well for billions of years on Earth.

BLACK SWAN TECHNOLOGIES

VINOD KHOSLA

Technology entrepreneur and venture capitalist, Khosla Ventures; formerly general partner at Kleiner Perkins Caufield & Byers; founder, Sun Microsystems

Think back to the world ten years ago. Google had just gotten started; Facebook and Twitter didn't exist. There were no smartphones; no one had remotely conceived of the possibility of the hundred thousand iPhone apps that exist today. The few large-impact technologies (versus slightly incremental advances in technologies) that have occurred in the past ten years are black-swan technologies. In his book *The Black Swan*, Nassim Taleb defines a black swan as an event of low probability, extreme impact, and only retrospective predictability. Black swans can be positive or negative in their effects and are found in every sector. Still, the most pressing reason I believe black-swan technology to be a conceptual tool that should be in everybody's cognitive toolkit is simply that the challenges of climate change and energy production we face are too big to be tackled by known solutions and safe bets.

I recall fifteen years ago, when we were starting Juniper networks, that there was absolutely no interest in replacing traditional telecommunications infrastructure (ATM was the mantra) with Internet protocols. After all, there were hundreds of billions of dollars invested in the legacy infrastructure, and it looked as immovable as today's energy infrastructure. Conventional wisdom would recommend incremental improvements to maximize the potential of the existing infrastructure. The fundamental flaw in conventional wisdom is the failure to acknowledge the possibility of a black swan. The likely future is not a traditional econometric forecast but rather one of today's improbables becoming tomorrow's conventional wisdom. Who would be crazy enough to have forecast in 2000 that by 2010 almost twice as many people in India would have access to cell phones as to latrines? Wireless phones were once only for the very rich. With a black-swan technology shot, you need not be constrained by the limits of the current infrastructure, projections, or market. You simply change the assumptions.

Many argue that since we already have some alternative energy technology, we should quickly deploy it. They fail to see the potential of black-swan technology possibilities; they discount them because they mistake "improbable" for "unimportant" and cannot imagine the art of the possible that technology enables; this alone runs the risk of spending vast amounts of money on outdated conventional wisdom. Even more important, it won't solve the problems we face. Focusing on short-term, incremental solutions will only distract us from working on producing the home runs that could change the assumptions regarding energy and society's resources. While there is no shortage of existing technology providing incremental improvements today (be it thin-film solar cells, wind turbines, or lithium-ion batteries), even summed they are simply irrelevant to the scale of our problems. They may make for interesting and sometimes large businesses but will not affect the prevailing energy and resource issues at scale. For that, we must look for and invest in quantum jumps in technology with low probability of success; we must create black-swan technologies. We must enable the multiplication of resources that only technology can create.

So, what are these next-generation technologies, these black-swan technologies of energy? They are risky investments that stand a high chance of failure but enable larger technological leaps that promise earth-shaking impact if successful: making solar power cheaper than coal, or viable without subsidies; economically making lighting and air conditioning 80 percent more efficient. Consider 100-percent-more-efficient vehicle engines, ultra-cheap energy storage, and countless other technological leaps we can't yet imagine. It's unlikely that any single shot works, of course. But even ten Google-like disruptions out of ten thousand shots will upend conventional wisdom, econometric forecasts, and, most important, our energy future.

To do so, we must reinvent the infrastructure of society by harnessing and motivating bright minds with a whole new set of future assumptions, asking "What could possibly be?" rather than "What is?" We need to create a dynamic environment of creative contention and collective brilliance that will yield innovative ideas from across disciplines to allow innovation to triumph. We must encourage a social ecosystem that encourages taking risks on innovation. Popularization of the concept of black-swan technology is essential to incorporate the right mind-set into entrepreneurs, policy makers, investors, and the public: that anything (maybe even everything) is possible. If we harness and motivate these bright new minds with the right market signals and encouragement, a whole new set of future assumptions, unimaginable today, will be tomorrow's conventional wisdom.

KAKONOMICS

GLORIA ORIGGI
Philosopher, Institut Jean Nicod, CNRS, Paris

An important concept explaining why life so often sucks is *kakonomics*, or the weird preference for low-quality payoffs.

Standard game-theoretical approaches posit that whatever people are trading (ideas, services, goods), each one wants to receive high-quality work from others. Let's stylize the situation so that goods can be exchanged only at two quality levels, high and low. Kakonomics (from the Greek, the economics of the worst) describes cases wherein people not only have the standard preference for receiving high-quality goods and delivering low-quality goods (the standard sucker's payoff) but actually *prefer* to deliver a low-quality product and receive a low-quality one: that is, they connive on a low-low exchange.

How can it be possible? And how can it be rational? Even when we're lazy and prefer to deliver a low-quality outcome (such as preferring to write a piece for a mediocre journal, provided they don't ask us to do too much work), we should still prefer to work less and receive more—that is, deliver low-quality and receive high-quality. Kakonomics is different: Here, we prefer not only to deliver a low-quality product but also to receive a low-quality good in exchange!

Kakonomics is the strange yet widespread preference for mediocre exchanges insofar as nobody complains about them. Kakonomic worlds are worlds in which people not only live with one another's laxness but expect it: I trust you not to keep your promises in full because I want to be free not to keep mine *and* not to feel bad about it. What makes it an interesting and weird case is that in all kakonomic exchanges, the two parties seem to have a double deal: an official pact in which both declare their intention to exchange at a high-quality level, and a tacit accord whereby discounts are not only allowed but expected. Thus, nobody is free-riding: Kakonomics is regulated by a tacit social norm of discount on quality, a mutual acceptance of a mediocre outcome, satisfactory to both parties as long as they aver publicly that the exchange is in fact at a high-quality level.

Take an example: A well-established best-selling author has to deliver his long overdue manuscript to his publisher. He has a large audience and knows very well that people will buy his book just because of his name—and, anyway, that the average reader doesn't read more than the first chapter. His publisher knows this as well. Thus, the author decides to deliver a manuscript with a stunning opening and a mediocre plot (the low-quality outcome). The publisher is happy with it and congratulates the author as though he'd delivered a masterpiece (the high-quality rhetoric), and both are satisfied. The author prefers not only to deliver a low-quality work but also that the publisher's response will be low-quality by failing to provide a serious edit and consenting to publish. They trust each other's untrustworthiness and connive on a mutually advantageous low outcome. Whenever there is a tacit deal to converge on low quality to mutual advantage, we're dealing with a case of kakonomics.

Paradoxically, if one of the two parties delivers a high-quality outcome instead of the expected low one, the other party resents it as a breach of trust, even if he may not acknowledge this openly. In the example, the author may resent the publisher if the latter delivers a high-quality edit. Trustworthiness in this relation would mean delivery of low quality too. Contrary to the standard game of the Prisoner's Dilemma, the willingness to repeat an interaction with someone is ensured if he or she delivers low quality too, rather than high quality.

Kakonomics is not always a bad thing. Sometimes it allows a certain discount that makes life more relaxing for everybody. As a friend who was renovating a country house in Tuscany told me: "Italian builders never deliver when they've promised to, but the good thing is, they don't expect you to pay them when you've promised to, either."

But the major problem of kakonomics and the reason it is a form of collective insanity so difficult to eradicate is that each low-quality exchange is a local equilibrium, in which both parties are satisfied; however, each of these exchanges erodes the overall system in the long run. So the threat to good collective outcomes doesn't come only from free riders and predators, as mainstream social sciences teach us, but also from well-organized norms of kakonomics, which regulate exchanges for the worse. The cement of society is not just cooperation for the good: In order to understand why life sucks, we should look also at norms of cooperation for a local optimum and an overall worsening.

KAYFABE

ERIC WEINSTEIN

Mathematician and economist; principal, Natron Group

The sophisticated scientific concept with the greatest potential to enhance human understanding may come not from academe but rather from the unlikely environment of professional wrestling.

Evolutionary biologists Richard Alexander and Robert Trivers have recently emphasized that deception rather than information often plays the decisive role in systems of selective pressures. Yet most of our thinking treats deception as a perturbation in the exchange of pure information, leaving us unprepared to contemplate a world in which fakery may reliably crowd out the genuine. In particular, humanity's future selective pressures appear likely to remain tied to economic theory that uses as its central construct a market model based on assumptions of perfect information.

If we are to take selection in humans more seriously, we may fairly ask what rigorous system could handle an altered reality of layered falsehoods, in which nothing can be assumed to be as it appears. Such a system, in development for more than a century, now supports an intricate multibillion-dollar business empire of pure hokum. It is known to wrestling's insiders as "kayfabe," a word of mysterious origin.

Because professional wrestling is a simulated sport, competitors who face each other in the ring are actually collaborators who must form a closed system (called "a promotion"), sealed against outsiders. Antagonists are chosen from within the promotion, and their ritualized battles are largely negotiated, choreographed, and rehearsed, at a significantly decreased risk of injury or death. With outcomes predetermined under kayfabe, betrayal in wrestling comes not from engaging in unsportsmanlike conduct but from the surprise appearance of actual sporting behavior. Such unwelcome sportsmanship, which "breaks kayfabe," is called "shooting" to distinguish it from the expected scripted deception, called "working."

Were kayfabe to become part of our toolkit for the twenty-first century, we would undoubtedly have an easier time understanding a world in which investigative journalism seems to have vanished and bitter corporate rivals cooperate on everything from joint ventures to lobbying efforts. Confusing battles between "freshwater" Chicago macroeconomists and Ivy League "saltwater" theorists could be best understood as happening within a single "orthodox promotion," given that both groups suffered no injury from failing (equally) to predict the recent financial crisis. The decades-old battle in theoretical physics over bragging rights between the string and loop camps seems an even more significant example within the hard sciences of a collaborative intrapromotional rivalry, given the apparent failure of both groups to produce a quantum theory of gravity.

What makes kayfabe remarkable is that it provides the most complete example of the process by which a wide class of important endeavors transition from failed reality to successful fakery. While most modern sports enthusiasts are aware of wrestling's status as a pseudosport, what few remember is that it evolved out of a failed real sport known as "catch" wrestling, which held its last honest title match early in the twentieth century. Typical matches could last hours with no satisfying action or end suddenly with crippling injuries to a promising athlete in whom much had been invested. This highlighted the close relationship between two paradoxical risks that define the category of activity that wrestling shares with other human spheres: (a) occasional but extreme peril for the participants and (b) general monotony for both audience and participants.

Kayfabrication (the process of transition from reality toward kayfabe) arises out of attempts to deliver a dependably engaging product for a mass audience while removing the unpredictable upheavals that imperil the participants. Thus, kayfabrication is a feature of many of our most important systems—such as war, finance, love, politics, and science. Importantly, kayfabe also illustrates the limits of disbelief the human mind is capable of successfully suspending before fantasy and reality become fully conflated. Wrestling's system of lies has recently become so intricate that wrestlers have occasionally found themselves engaging in real-life adultery following the introduction of a fictitious adulterous plot twist in a kayfabe backstory. Eventually, even kayfabe itself became a victim of its own success, as it grew to a level of deceit that could not be maintained when the wrestling world collided with outside regulators exercising oversight over major sporting events.

When kayfabe was forced to own up to the fact that professional wrestling contained no sport whatsoever, it did more than avoid being regulated and taxed into oblivion. Wrestling discovered the unthinkable: Its audience did not seem to require even a thin veneer of realism. Professional wrestling had come full circle to its honest origins by at last moving the responsibility for deception off the shoulders of the performers and into the willing minds of the audience.

Kayfabe, it appears, is a dish best served clientside.

EINSTEIN'S BLADE IN OCKHAM'S RAZOR

KAI KRAUSE
Software pioneer; interface designer

In 1971, when I was a teenager, my father died in an airplane crash. Somehow I began to turn "serious," trying to understand the world around me and my place in it, looking for meaning and sense, beginning to realize that everything was different from what I had assumed in the innocence of childhood.

It was the beginning of my own "building a cognitive toolkit," and I remember the pure joy of discovery, reading voraciously and—quite out of sync with friends and school—devouring encyclopedias, philosophy, biographies, and science fiction.

One such story stayed with me, and one paragraph within it especially: "*We need to make use of Thargola's Sword! The principle of Parsimony. First put forth by the medieval philosopher Thargola14, who said, 'We must drive a sword through any hypothesis that is not strictly necessary.'* "

That really made me think—and think again.

Finding out who this man might have been took quite a while, but it was also another beginning: a love affair with libraries, large tomes, dusty bindings . . . surfing *knowledge*, as it were. And I did discover that there had been a monk, from a hamlet surrounded by oaks, apocryphally named William of Ockham. He crossed my path again years later, when I was lecturing in Munich, near Occam Street, and realized that he had spent the last twenty years of his life there, under King Ludwig IV, in the mid-1300s.

Isaac Asimov had pilfered, or let's say homaged, good old William for what is now known in many variants as "Ockham's razor," such as

Plurality should not be posited without necessity.

Entities are not to be multiplied beyond necessity.

Or more general and colloquial, and a bit less transliterated from Latin:

A simpler explanation invoking fewer hypothetical constructs is preferable.

And there it was, the dancing interplay between *simplex and complex* that has fascinated me in so many forms ever since. For me, it is very near the center of "understanding the world," as our question posited.

Could it really be true that the innocent-sounding "Keep it simple" is really such an optimal strategy for dealing with questions large and small, scientific as well as personal? Surely, trying to eliminate superfluous assumptions can be a useful tenet and can be found from Sagan to Hawking as part of their approach to thinking in science. But something never quite felt right to me. Intuitively, it was clear that sometimes things are just not simple—and that merely "the simplest" of all explanations cannot be taken as truth or proof.

- Any detective story would pride itself in not using the most obvious explanation of who did it or how it happened.
- Designing a car to have "the optimal feel going into a curve at high speed" would require hugely complex systems to finally arrive at "simply good."
- Water running downhill will take a meandering path instead of a straight line.

The non-simple solution is "the easiest," seen from another viewpoint: For the water, the *least energy used going down the shal-*

lowest slope is more important than *taking the straightest line from A to B*. And that is one of the issues with Ockham: The definition of what "simple" is can already be anything but simple. And what "simpler" is—well, it just doesn't get any simpler there.

There is that big difference between *simple* and *simplistic*. And seen more abstractly, the principle of *simple things leading to complexity* dances in parallel and involved me deeply throughout my life.

In the early seventies I also began tinkering with the first large-scale modular synthesizers, finding quickly how hard it is to re-create seemingly "simple" sounds.

There was unexpected complexity in a single note struck on a piano—complexity that eluded even dozens of oscillators and filters by magnitudes.

Lately, one of many projects has been to revisit the aesthetic space of scientific visualizations, and another the epitome of mathematics made tangible: fractals, which I had done almost twenty years ago with virtuoso coder Ben Weiss, now enjoying them via *realtime flythroughs* on a handheld little smartphone. Here was the most extreme example: A tiny formula, barely one line on paper, used recursively, yields *worlds* of complex images of amazing beauty. (Ben had the distinct pleasure of showing Benoit Mandelbrot an alpha version at a TED conference just months before Mandelbrot's death.)

My hesitation about overuse of parsimony was expressed perfectly in a quote from Albert Einstein, arguably the counterpart blade to Ockham's razor: "*Things should be made as simple as possible—but not simpler.*"

And there we have the perfect application of its truth, used recursively on itself: *Neither Einstein nor Ockham actually used the exact words as quoted!* After I sifted through dozens of books, his collected works and letters in German, the Einstein Archives: Nowhere there, nor in Britannica, Wikipedia, or Wikiquote, was anyone able to substantiate exact sources. And the same applies to Ockham. If anything *can* be found, it is precedences.

Surely, one can amass retweeted, reblogged, and regurgitated instances for both very quickly—they have become memes, of course. One could also take the view that in each case they certainly "might" have said it "just like that," since each used several expressions similar in form and spirit. But to attribute the exact words because they're "kind of close" would be—well, another case of *it's not that simple*!

And there is a huge difference between *additional* and *redundant* information. (Or else one could lose the second, redundant "ein" in "Einstein"?)

Linguistic jesting aside: The Razor and the Blade constitute a useful combined approach to analytical thinking. Shaving away nonessential conjectures is a good thing, a worthy inclusion in "everybody's toolkit"—and so is the corollary: *Don't overdo it!*

And my own bottom line: *There is nothing more complex than simplicity.*

HEAT-SEEKING MISSILES

DAVE WINER

Visiting scholar in journalism, New York University; pioneer software developer (blogging, podcasting, RSS, outliners, Web content management)

New York City, my new home, teaches you that although we are social creatures, it's often best not to admit it.

As you weave among the obstacles on the sidewalks of Manhattan, it's easy to get distracted from your thoughts and pay attention to the people you're encountering. It's OK to do that if you're at a halt, but if you're in motion and your eyes engage with another, that signals you would like to negotiate.

Not good. A sign of weakness. Whether the oncoming traffic is aware or not, he or she will take advantage of this weakness and charge right into your path, all the while not making eye contact. There is no appeal. All you can do is shift out of his or her path, but even this won't avoid a collision, because your adversary will unconsciously shift closer to you. Your weakness is attractive. Your space is up for grabs. At this point, you have no choice but to collide, and in the etiquette of New York street walking, you're responsible.

That's why the people who check their smartphones for text messages or e-mails while walking so totally command the sidewalks. They are heat-seeking missiles, and it's your heat they seek.

I don't think this is just New York; it's a feature of the human species. We seek companionship.

For a while in 2005 I lived on the beach in northeast Florida outside St. Augustine. The beach is long and relatively empty; they let you drive on the beach to find the perfect spot to bathe, and if you're willing to drive a bit, you can be alone. So I would drive to a secluded spot, park my car, and go out into the surf. When I came back, more often than not, there was a car parked right next to mine. They could have parked anywhere within a mile in either direction and had it all to themselves.

Add that to your cognitive toolkit!

ENTANGLEMENT

MARCO IACOBONI
Neuroscientist; professor of psychiatry and biobehavioral sciences; director of the Transcranial Magnetic Stimulation Lab, Ahmanson-Lovelace Brain Mapping Center, David Geffen School of Medicine, University of California–Los Angeles; author, Mirroring People

Entanglement is "spooky action at a distance," as Einstein liked to say (he actually did not like it at all, but at some point he had to admit that it exists). In quantum physics, two particles are entangled when a change in one particle is immediately associated with a change in the other particle. Here comes the spooky part: We can separate our "entangled buddies" as far as we can, and they will remain entangled. A change in one is instantly reflected in the other, even though they are physically far apart (and I mean in different countries!).

Entanglement feels like magic. It is really difficult to wrap our heads around it. Yet entanglement is a real phenomenon, measurable and reproducible in the lab. And there is more. While for many years entanglement was thought to be a very delicate phenomenon, observable only in the infinitesimally small world of quantum physics ("Oh good, our world is immune from that weird stuff") and quite volatile, recent evidence suggests that entanglement may be much more robust and widespread than we initially thought. Photosynthesis may happen through entanglement, and recent brain data suggest that entanglement may play a role in coherent electrical activity of distant groups of neurons in the brain.

Entanglement is a good cognitive chunk, because it challenges our cognitive intuitions. Our minds seem built to prefer relatively mechanical cause-and-effect stories as explanations of natural phenomena. And when we can't come up with one of those stories, we tend to resort to irrational thinking—the kind of magic we feel when we think about entanglement. Entangled particles teach us that our beliefs about how the world works can seriously interfere with our understanding of it. But they also teach us that if we stick with the principles of good scientific practice, of observing, measuring, and then reproducing phenomena that we can frame in a theory (or that are predicted by a scientific theory), we can make sense of things. Even weird things, like entanglement.

Entanglement is also a good cognitive chunk because it whispers to us that seemingly self-evident cause-and-effect phenomena may not be cause-and-effect at all. The timetable of modern vaccination, probably the biggest accomplishment in modern medicine, coincides with the onset of symptoms of autism in children. This temporal correspondence may mislead us to think that the vaccination may have produced the symptoms, hence the condition of autism. At the same time, that temporal correspondence should make us suspicious of straightforward cause-and-effect associations, inviting us to take a second look and conduct controlled experiments to find out whether or not there really is a link between vaccines and autism. We now know there is no such link. Unfortunately, this belief is hard to eradicate and is producing in some parents the potentially disastrous decision to not vaccinate their children.

The story of entanglement is a great example of the capacity of the human mind for reaching out almost beyond itself. The key word here is "almost." Because we "got there," it is self-evident that we could "get there." But it didn't feel like it, did it? Until we managed to observe, measure, and reproduce that phenomenon predicted by quantum theory, it just felt a little "spooky." (It still feels a bit spooky, doesn't it?) Humans are naturally inclined to reject facts that do not fit their beliefs—and, indeed, when confronted with those facts they tend to automatically reinforce their beliefs and brush the facts under the carpet. The beautiful story of entanglement reminds us that we can go "beyond ourselves," that we don't have to desperately cling to our beliefs, and that we can make sense of things. Even spooky ones.

TECHNOLOGY PAVED THE WAY FOR HUMANITY

TIMOTHY TAYLOR
Archaeologist, University of Bradford, UK; author, The Artificial Ape: How Technology Changed the Course of Human Evolution

The very idea of a "cognitive toolkit" is one of the most important items in our cognitive toolkit. It is far more than just a metaphor, for the relationship between actual physical tools and the way we think is profound and of immense antiquity.

Ideas such as evolution and a deep prehistory for humanity are as factually well established as the idea of a round Earth. Only bigots and the misled can doubt them. But the idea that the first chipped stone tool predates, by at least half a million years, the expansion of mind that is so characteristic of humans should also be knowable by all.

The idea that technology came before humanity and, evolutionarily, paved the way for it, is the scientific concept that I believe should be part of everybody's cognitive toolkit. We could then see that thinking through things and with things, and manipulating virtual things in our minds, is an essential part of critical self-consciousness. The ability to internalize our own creations by abstracting them and converting "out there" tools into mental mechanisms is what allows the entire scientific project.

TIME SPAN OF DISCRETION

PAUL SAFFO
Technology forecaster; managing director of foresight at Discern Analytics; distinguished visiting scholar in the Stanford Media X research network, Stanford University

Half a century ago, while advising a UK Metals company, Elliott Jaques had a deep and controversial insight. He noticed that workers at different levels of the company had very different time horizons. Line workers focused on tasks that could be completed in a single shift, whereas managers devoted their energies to tasks requiring six months or more to complete. Meanwhile, their CEO was pursuing goals realizable only over the span of several years.

After several decades of empirical study, Jaques concluded that just as humans differ in intelligence, we differ in our ability to handle time-dependent complexity. We all have a natural time horizon we are comfortable with: what Jaques called "time span of discretion," or the length of the longest task an individual can successfully undertake. Jaques observed that organizations implicitly recognize this fact in everything from titles to salary. Line workers are paid hourly, managers annually, and senior executives compensated with longer-term incentives, such as stock options.

Jaques also noted that effective organizations were comprised of workers of differing time spans of discretion, each working at a level of natural comfort. If a worker's job was beyond his natural time span of discretion, he would fail. If it was less, he would be insufficiently challenged and thus unhappy.

Time span of discretion is about achieving intents that have explicit time frames. And in Jaques's model, one can rank discretionary capacity in a tiered system. Level 1 encompasses jobs such as sales associates or line workers, handling routine tasks with a time horizon of up to three months. Levels 2 to 4 encompass various managerial positions, with time horizons from one to five years. Level 5 encompasses five to ten years and is the domain of small-company CEOs and large-company executive vice presidents. Beyond Level 5, one enters the realm of statesmen and legendary business leaders, comfortable with innate time horizons of twenty years (Level 6), fifty years (Level 7) or beyond. Level 8 is the realm of hundred-year thinkers, like Henry Ford, while Level 9 is the domain of the Einsteins, Gandhis, and Galileos, individuals capable of setting grand tasks into motion that continue centuries into the future.

Jaques's ideas enjoyed currency into the 1970s and then fell into eclipse, assailed as unfair stereotyping or worse, a totalitarian stratification evocative of Aldous Huxley's *Brave New World.* It is now time to reexamine Jaques's theories and revive "time span of discretion" as a tool for understanding our social structures and matching them to the overwhelming challenges facing global society. Perhaps problems like climate change are intractable because we have a political system that elects Level 2 thinkers to Congress, when we really need Level 5's in office. As such, Jaques's ideas might help us realize that the old saying "He who thinks longest wins" is only half the story, and that the society in which everyone explicitly thinks about tasks in the context of time will be the most effective.

DEFEASIBILITY

TANIA LOMBROZO
Cognitive psychologist, University of California–Berkeley

On its face, defeasibility is a modest concept, with roots in logic and epistemology. An inference is defeasible if it can potentially be "defeated" in light of additional information. Unlike deductively sound conclusions, the products of defeasible reasoning remain subject to revision, held tentatively no matter how firmly.

All scientific claims—whether textbook pronouncements or haphazard speculations—are held defeasibly. It is a hallmark of the scientific process that claims are forever vulnerable to refinement and rejection, hostage to what the future could bring. Far from being a weakness, this is a source of science's greatness. Because scientific inferences are defeasible, they remain responsive to a world that can reveal itself gradually, change over time, and deviate from our dearest assumptions.

The concept of defeasibilility has proved valuable in characterizing artificial and natural intelligence. Everyday inferences, no less than scientific inferences, are vetted by the harsh judge of novel data—additional information that can potentially defeat current beliefs. On further inspection, the antique may turn out to be a fake and the alleged culprit an innocent victim. Dealing with an uncertain world forces cognitive systems to abandon the comforts of deduction and engage in defeasible reasoning.

Defeasibility is a powerful concept when we recognize it not as a modest term of art but as the proper attitude toward all belief. Between blind faith and radical skepticism is a vast but sparsely populated space where defeasibility finds its home. Irreversible commitments would be foolish, boundless doubt paralyzing.

Defeasible beliefs provide the provisional certainty necessary to navigate an uncertain world.

Recognizing the potential revisability of our beliefs is a prerequisite to rational discourse and progress, be it in science, politics, religion, or the mundane negotiations of daily life. Consider the world we could live in if all of our local and global leaders, if all of our personal and professional friends and foes, recognized the defeasibility of their beliefs and acted accordingly. That sure sounds likc progress to me. But of course I could be wrong.

AETHER

RICHARD THALER
Economist; director, Center for Decision Research, Booth School of Business, University of Chicago; coauthor (with Cass Sunstein), Nudge: Improving Decisions About Health, Wealth, and Happiness

I recently posted a question on *Edge* asking people to name their favorite example of a wrong scientific belief. One of my prized answers came from Clay Shirky. Here is an excerpt:

> The existence of ether, the medium through which light (was thought to) travel. It was believed to be true by analogy—waves propagate through water, and sound waves propagate through air, so light must propagate through X, and the name of this particular X was ether.
>
> It's also my favorite because it illustrates how hard it is to accumulate evidence for deciding something doesn't exist. Ether was both required by nineteenth-century theories and undetectable by nineteenth-century apparatus, so it accumulated a raft of negative characteristics: it was odorless, colorless, inert, and so on.

Several other entries (such as the "force of gravity") shared the primary function of ether: They were convenient fictions able to "explain" some otherwise ornery facts. Consider this quote from Max Pettenkofer, the German chemist and physician, disputing the role of bacteria as a cause of cholera: "Germs are of no account in cholera! The important thing is the disposition of the individual."

So in answer to the current *Edge* Question, I am proposing that we now change the usage of the word "Aether," using the old spelling, since there is no need for a term that refers to something that does not exist. Instead, I suggest we use that term to describe the role of any free parameter used in a similar way: that is, "Aether is the thing that makes my theory work." Replace the word "disposition" with "Aether" in Pettenkofer's sentence to see how it works.

Often, Aetherists (theorists who rely on an Aether variable) think their use of the Aether concept renders their theory untestable. This belief is often justified during their lifetimes, but then along come clever empiricists, such as A. A. Michelson and Edward Morley, and last year's tautology becomes this year's example of a wrong theory.

Aether variables are extremely common in my own field of economics. Utility is the thing you must be maximizing in order to render your choice rational.

Both risk and risk aversion are concepts that were once well defined but are now in danger of becoming Aetherized. Stocks that earn surprisingly high returns are labeled as risky, because, in the theory, excess returns must be accompanied by higher risk. If, inconveniently, the traditional measures of risk, such as variance or covariance with the market, are not high, then the Aetherists tell us there must be some other risk; we just don't know what it is.

Similarly, traditionally the concept of risk aversion was taken to be a primitive; each person had a parameter, *gamma*, that measured her degree of risk aversion. Now risk aversion is allowed to be time varying, and Aetherists can say with straight faces that the market crashes of 2001 and 2008 were caused by sudden increases in risk aversion. (Note the direction of the causation. Stocks fell because risk aversion spiked, not vice versa.)

So the next time you are confronted with such a theory, I suggest substituting the word "Aether" for the offending concept. Personally, I am planning to refer to the time-varying variety of risk aversion as Aether aversion.

KNOWLEDGE AS A HYPOTHESIS

MARK PAGEL

Professor of evolutionary biology, University of Reading, UK; external professor, Santa Fe Institute

The Oracle of Delphi famously pronounced Socrates to be "the most intelligent man in the world because he knew that he knew nothing." Over two thousand years later, the mathematician-turned-historian Jacob Bronowski would emphasize—in the last episode of his landmark 1970s television series *The Ascent of Man*—the danger of our all-too-human conceit of thinking we know something, as evidenced in the Nazi atrocities of the Second World War. What Socrates knew, and what Bronowski had come to appreciate, is that knowledge—true knowledge—is difficult, maybe even impossible, to come by. It is prone to misunderstanding and counterfactuals, and, most important, it can never be acquired with exact precision; there will always be some element of doubt about anything we come to "know" from our observations of the world.

What is it that adds doubt to our knowledge? It is not just the complexity of life; uncertainty is built into anything we measure. No matter how well you can measure something, you might be wrong by up to half the smallest unit you can discern.

If you tell me I am 6 feet tall, and you can measure to the nearest inch, I might actually be 5' 11 ½" or 6' ½" and you (and I) won't know the difference. If something is really small, you won't even be able to measure it, and if it is really, really small, a light microscope (and thus your eye, both of which can see only objects larger than the shortest wavelength of visible light) won't even know it's there.

What if you measure something repeatedly?

This helps, but consider the plight of those charged with international standards of weights and measures. There is a lump of metal stored under a glass case in Sèvres, France. It is, by the decree of Le Système Internationale d'Unités, the definition of a kilogram. How much does it weigh? Well, by definition, whatever it weighs is a kilogram. But the fascinating thing is that it has never *weighed* exactly the same twice. On those days that it weighs less than a kilogram, you're not getting such a good deal at the grocery store. On other days, you are.

The often blithe way in which scientific "findings" are reported by the popular press can mask just how difficult it is to acquire reliable knowledge. Height and weight are—as far as we know—single dimensions. Consider, then, how much more difficult it is to measure something like intelligence, or the risk of getting cancer from eating too much meat, or whether cannabis should be legalized, or whether the climate is warming and why, or what a "shorthand abstraction" or even "science" is, or the risk of developing psychosis from drug abuse, or the best way to lose weight, or whether it is better to force people receiving state benefits to work, or whether prisons are a deterrent to crime, or how to quit smoking, or whether a glass of wine every day is good for you, or whether 3-D glasses will hurt your children's eyes, or even just the best way to brush your teeth. In each case, what was actually measured, or who was measured? Who were they compared to, for how long? Are they like you and me? Were there other factors that could explain the outcome?

The elusive nature of knowledge should remind us to be humble when interpreting it and acting on it, and this should grant us both a tolerance and skepticism toward others and their interpretations. Knowledge should always be treated as a hypothesis. It has only just recently emerged that Bronowski, whose family was slaughtered at Auschwitz, himself worked with Britain's Royal Air Force during the Second World War calculating how best to deliver bombs—vicious projectiles of death that don't discriminate between good guys and bad guys—to the cities of the Third Reich. Maybe Bronowski's later humility was born of this realization—that our views can be wrong and they can have consequences for others' lives. Eager detractors of science as a way of understanding the world will jump on these ideas with glee, waving them about as proof that "nothing is real" and that science and its outputs are as much a human construct as art or religion. This is facile, ignorant, and naïve.

Measurement and the "science" or theories it spawns must be treated with humility precisely because they are powerful ways of understanding and manipulating the world. Observations can be replicated—even if imperfectly—and others can agree on how to make the measurements on which they depend, be they measurements of intelligence, the mass of the Higgs boson, poverty, the speed at which proteins can fold into their three-dimensional structures, or how big gorillas are.

No other system for acquiring knowledge even comes close to science, but this is precisely why we must treat its conclusions with humility. Einstein knew this when he said that "all our science measured against reality is primitive and childlike, and yet," he added, "it is the most precious thing we have."

THE EINSTELLUNG EFFECT

EVGENY MOROZOV
Commentator on Internet and politics, Net Effect blog; contributing editor, Foreign Policy; *author,* The Net Delusion: The Dark Side of Internet Freedom

Constant awareness of the Einstellung effect would make a useful addition to our cognitive toolkit. The Einstellung effect is more pervasive than its name suggests. We constantly experience it when trying to solve a problem by pursuing solutions that have worked for us in the past, instead of evaluating and addressing the new problem on its own terms. Thus, whereas we may eventually solve the problem, we may be wasting an opportunity to do so in a more rapid, effective, and resourceful manner.

Think of a chess match. If you are a chess master with a deep familiarity with chess history, you are likely to spot game developments that look similar to other matches you know by heart. Knowing how those previous matches unfolded, you may automatically pursue similar solutions.

This may be the right thing to do in matches that are exactly alike, but in other situations you've got to watch out! Familiar solutions may not be optimal. Recent research into the occurrences of the Einstellung effect in chess players suggests that it tends to be less prominent once they reach a certain level of mastery; they get a better grasp of the risks associated with pursuing solutions that look familiar, and they avoid acting on "autopilot."

The irony here is that the more expansive our cognitive toolkit, the more likely we are to fall back on past solutions and approaches instead of asking whether the problem in front of us is fundamentally different from anything else we have dealt with. A cognitive toolkit that has no built-in awareness of the Einstellung effect seems defective to me.

HOMO SENSUS SAPIENS: THE ANIMAL THAT FEELS AND REASONS

EDUARDO SALCEDO-ALBARÁN

Philosopher; founder and manager, Metodo, *a transdisciplinary, transnational group of social scientists*

For the last three years, Mexican narcotraffickers have decapitated hundreds of people to gain control of routes for transporting cocaine. In the last two decades, Colombian narcoparamilitaries tortured and incinerated thousands of people, in part because they needed more land for their crops and for transporting cocaine. In both cases, the perpetrators were not satisfied with $10 million or $100 million; even the richest narcotraffickers kill, or die, for more.

In Guatemala and Honduras, vicious battles between gangs known as *maras* are waged to gain control of a street in a poor neighborhood. In Rwanda's genocide, in 1994, people who had been friends all their lives suddenly became mortal enemies because of ethnicity.

Is this enlightened?

These cases may seem like rarities. However, in any city, in any random street, it is easy to find a thief willing to kill or die for ten bucks to satisfy the need for heroin; a fanatic willing to kill or die in defense of a "merciful God"; a regular guy-next-door willing to kill or die in a fight after a car crash.

Is this rational?

Examples abound in which such automatic emotional responses as ambition, anger, or anxiety overcome rationality. Those responses keep assaulting us, like uncontrollable forces of nature—like storms, or earthquakes.

We modern humans taxonomically define ourselves as *Homo sapiens sapiens*, wise wise beings. Apparently we can dominate natural forces, be they instincts, viruses, or storms. However, we cannot avoid destroying natural resources while consuming more than we need. We cannot control excessive ambition. We cannot avoid surrendering to the power of sex or money. Despite our evolved brains, despite our ability to argue and think in abstract ways, despite the amazing power of our neocortex, our innermost feelings are still at the base of our behavior.

Neurological observations indicate that instinctive areas of the brain are active most of the time. Our nervous system is constantly at the mercy of neurotransmitters and hormones that determine levels of emotional responses. Observations from experimental psychology and behavioral economics show that people do not always try to maximize present or future profits. Rational expectations, once thought of as the main characteristic of *Homo economicus*, are not neurologically sustainable anymore. Sometimes people want only to satisfy a desire right here, right now, no matter what.

Human beings do have unique rational capacities. No other animal can evaluate, simulate, and decide for the best as humans do; however, having the capacity doesn't always mean executing it.

The inner and oldest areas of the human brain—the reptilian brain—generate and regulate instinctive and automatic responses, which play a role in preserving the organism. Because of these areas, we move without analyzing the consequence of each action; we move like a machine of automatic and unconscious induction. We walk without determining whether the floor will remain solid after each step. We run when we feel a threat, not because of rational planning but automatically.

Only strict training enables us to dominate our instincts. For most of us, the admonition "Don't panic" works only when we're not panicking. Most of us should be defined as beings moved initially by instincts, empathy, and automatic responses resulting from our perceptions, instead of by sophisticated plans and arguments. *Homo economicus* and *Homo politicus* are behavioral benchmarks rather than descriptive models. The calculation of utility and the resolution of social disputes through civilized debate are behavioral utopias, not descriptions of what we are. However, for decades we've been constructing policies, models, and sciences, not coinciding with reality, based on these assumptions. *Homo sensus sapiens* is a more accurate image of the human being.

The liberal hyperrationalist and the conservative hypercommunitarian are hypertrophies of a single human facet. The first is the hypertrophy of the neocortex: the idea that rationality dominates instincts. The second is the hypertrophy of the inner reptilian brain: the idea that empathy and cohesive institutions define humanity. However, we are both at the same time. We are the tension of the *sensus* and the *sapiens.*

The concept of *Homo sensus sapiens* allows us to realize that we are at a point somewhere between overconfidence in our rational capacities and submission to our instincts. It also allows us to improve our explanations of social phenomena. Social scientists should not always discriminate between rationality and irrationality. They should get out of the comfort zone of positivist fragmentation and integrate scientific areas to explain an analog human being, not a digital one—a human being defined by the continuum between sensitivity and rationality. Better inputs for public policy would be proposed with this adjusted image.

The first character of this *Homo*, the *sensus*, allows movement, reproduction, preservation of the species. The *sapiens* allows psychological oscillation between the ontological world of matter and energy and the epistemological world of sociocultural codification, imagination, arts, technology, and symbolic construction.

This combination allows understanding of the nature of a hominid characterized by the constant tension between emotions and reason, and the search for a middle point of biological and cultural evolution. We are not only fearers, not only planners. We are *Homo sensus sapiens*, the animal that feels and reasons.

UNDERSTANDING CONFABULATION

FIERY CUSHMAN

Assistant professor, Department of Cognitive, Linguistic & Psychological Sciences, Brown University

We are shockingly ignorant of the causes of our own behavior. The explanations we provide are sometimes wholly fabricated and certainly never complete. Yet that is not how it feels; instead, it feels as though we know exactly what we're doing and why. This is confabulation: guessing at plausible explanations for our behavior and then regarding those guesses as introspective certainties. Psychologists use dramatic examples to entertain their undergraduate audiences. Confabulation is funny, but there is a serious side, too. Understanding it can help us act and think better in everyday life.

Some of the most famous examples of confabulation come from split-brain patients, whose left and right brain hemispheres have been surgically disconnected for medical treatment. Neuroscientists have devised clever experiments in which information is provided to the right hemisphere (for instance, pictures of naked people), causing a change in behavior (embarrassed giggling). Split-brain individuals are then asked to explain their behavior verbally, which relies on the left hemisphere. Realizing that the body is laughing, but unaware of the nude images, the left hemisphere will confabulate an excuse for the body's behavior ("I keep laughing because you ask such funny questions, Doc!").

Wholesale confabulations in neurological patients can be jaw-dropping, but in part that's because they don't reflect ordinary experience. Most of the behaviors you and I perform are not induced by crafty neuroscientists planting subliminal suggestions in our right hemisphere. When we're outside the laboratory—

and when our brains have all the usual connections—most of our behaviors are the product of some combination of deliberate thinking and automatic action.

Ironically, that's exactly what makes confabulation so dangerous. If we routinely got the explanation for our behavior totally wrong—as completely wrong as split-brain patients sometimes do—we would probably be much more aware that there are pervasive, unseen influences on it. The problem is that we get all of our explanations partly right, correctly identifying the conscious and deliberate causes of our behavior. Unfortunately, we mistake "partly right" for "completely right," and thereby fail to recognize, and guard against, the equal influence of the unconscious.

A choice of job, for instance, depends partly on careful deliberation about career interests, location, income, and hours. At the same time, research reveals that choice to be influenced by a host of factors of which we are unaware. According to a 2005 study, people named Dennis or Denise are more likely to be dentists, while people named Virginia are more likely to locate to (you guessed it) Virginia [Pelham, Carvallo & Jones, *Psychol. Sci.*]. Less endearingly, research suggests that, on average, people will take a job with fewer benefits, a less desirable location, and a smaller income if it allows them to avoid having a female boss [Rahnev, Caruso & Banaji, 2007, unpub. ms., Harvard Univ.]. Surely, most people do not want to choose a job based on the sound of their name, nor do they want to sacrifice job quality in order to perpetuate old gender norms. Indeed, most people have no awareness that these factors influence their choices. When you ask them why they took the job, they are likely to reference their conscious thought processes: "I've always loved making ravioli, the *lira* is on the rebound, and Rome is for lovers . . ." That answer is partly right, but it is also partly wrong, because it misses the deep reach of automatic processes on human behavior.

People make harsher moral judgments in foul-smelling rooms, reflecting the role of disgust as a moral emotion [Schnall et al., 2008, *Pers. & Soc. Psych. Bull.*]. Women are less likely to call their fathers during the fertile phase of their menstrual cycle than during their non-fertile phase, reflecting a means of incest avoidance; no such pattern is found in calls to their mothers [Lieberman, Pillsworth & Haselton, 2010, *Psychol. Sci.*]. Students indicate greater political conservatism when polled near a hand-sanitizing station during a flu epidemic, reflecting the influence of a threatening environment on ideology [Helzer & Pizarro, 2011, *Psychol. Sci.*]. They will also judge a stranger to be more generous and caring when they hold hot coffee versus iced coffee, reflecting the metaphor of a "warm" relationship [Williams & Bargh, 2008, *Science*].

Automatic behaviors can be remarkably organized and even goal-driven. For example, research shows that people tend to cheat only as much as they can without realizing they're cheating [Mazar, Amir & Ariely, 2008, *Jour. Marketing Res.*]. This is a remarkable phenomenon: Part of you is deciding how much to cheat, calibrated at just the level that keeps another part of you from realizing it.

One of the ways people pull off this trick is with innocent confabulations: When self-grading an exam, students think, "Oh, I was going to circle *e*, I really knew that was the answer!" This isn't a lie, any more than it's a lie to say you don't have time to call your dad during this busy time of the month. These are just incomplete explanations, confabulations that reflect our conscious thoughts while ignoring the unconscious ones.

This brings me to the central point, the part that makes confabulation an important concept in ordinary life and not just a trick pony for college lectures. Perhaps you've noticed that people have an easier time sniffing out unseemly motivations for others' behavior than recognizing the same motivations for their own behavior. Others avoided female bosses (sexists) and inflated their grades (cheaters), while we chose Rome and really meant to say that Anne was the third Brontë. There's a double tragedy in this double standard.

First, we jump to the conclusion that others' behaviors reflect their bad motives and poor judgment, attributing conscious choice to behaviors that may have been influenced unconsciously. Second, we assume that our own choices were guided solely by the conscious explanations we conjure, and we reject or ignore the possibility that we may have unconscious biases of our own.

By understanding confabulation, we can begin to remedy both faults. We can hold others responsible for their behavior without necessarily impugning their conscious motivations. And we can hold ourselves more responsible by inspecting our own behavior for its unconscious influences, as unseen as they are unwanted.

SEXUAL SELECTION

DAVID M. BUSS
Professor of psychology, University of Texas–Austin; author, The Evolution of Desire: Strategies of Human Mating; *coauthor (with Cindy M. Meston),* Why Women Have Sex

When most people think about evolution by selection, they conjure up phrases such as "survival of the fittest" or "nature red in tooth and claw." These focus attention on the Darwinian struggle for survival. Many scientists, but few others, know that evolution by selection occurs through the process of differential *reproductive* success by virtue of heritable differences in design, not by differential survival success. And differential reproductive success often boils down to differential mating success, the focus of Darwin's 1871 theory of sexual selection.

Darwin identified two separate (but potentially related) causal processes by which sexual selection occurs. The first, intrasexual or same-sex competition, involves members of one sex competing with one another in various contests, physical or otherwise, whose winners gain preferential access to mates. Qualities that lead to success evolve; those linked to failure bite the evolutionary dust. Evolution, change over time, occurs as a consequence of the process of intrasexual competition. The second, intersexual selection, deals with preferential mate choice. If members of one sex exhibit a consensus about qualities desired in mates, and those qualities are partially heritable, then those of the opposite sex possessing the desired qualities have a mating advantage. They get preferentially chosen. Those lacking desired mating qualities get shunned, banished, and remain mateless (or must settle for low-quality mates). Evolutionary change over time occurs as a consequence of an increase in frequency of desired traits and a decrease in frequency of disfavored traits.

Darwin's theory of sexual selection, controversial in his day and relatively neglected for nearly a century after its publication, has mushroomed into a tremendously important theory in evolutionary biology and evolutionary psychology. Research on human mating strategies has exploded over the past decade, as the profound implications of sexual selection become more deeply understood. Adding sexual selection to everybody's cognitive toolkit will provide profound insights into many human phenomena that otherwise remain baffling. In its modern formulations, sexual-selection theory offers answers to weighty and troubling questions that still elude many scientists and most nonscientists:

- Why do male and female minds differ?
- What explains the rich menu of human mating strategies?
- Why is conflict between the sexes so pervasive?
- Why does conflict between women and men focus so heavily on sex?
- What explains sexual harassment and sexual coercion?
- Why do men die earlier than women, on average, in every culture around the world?
- Why are most murderers men?
- Why are men so much keener than women on forming coalitions for warfare?
- Why are men so much more prone to becoming suicide terrorists?
- Why is suicide terrorism so much more prevalent in polygynous cultures that create a greater pool of mateless males?

Adding sexual-selection theory to everybody's cognitive toolkit, in short, provides deep insight into the nature of human nature, our obsession with sex and mating, the origins of sex differences, and many of the profound social conflicts that beset us all.

QED MOMENTS

BART KOSKO

Professor of electrical engineering, University of Southern California; author, Noise

Everyone should know what proof feels like. It reduces all other species of belief to a distant second-class status. Proof is the far end on a cognitive scale of confidence that varies through levels of doubt. And most people never experience it.

Feeling proof comes from finishing a proof. It does not come from pointing at a proof in a book or in the brain of an instructor. It comes when the prover himself takes the last logical step on the deductive staircase. Then he gets to celebrate that logical feat by declaring "QED" or "*Quod erat demonstrandum*" or just "Quite easily done." QED states that he has proved or demonstrated the claim he wanted to prove. The proof need not be original or surprising. It just needs to be logically correct to produce a QED moment. A proof of the Pythagorean theorem has always sufficed.

The only such proofs that warrant the name are those in mathematics and formal logic. Each logical step has to come with a logically sufficient justification. That way each logical step comes with binary certainty. Then the final result itself follows with binary certainty. It is as if the prover had multiplied the number 1 by itself for each step in the proof. The result is still the number 1. That is why the final result warrants a declaration of QED. That is also why the process comes to an unequivocal halt if the prover cannot justify a step. Any act of faith or guesswork or cutting corners will destroy the proof and its demand for binary certainty.

The catch is that we can really only prove tautologies.

The great binary truths of mathematics are still logically equivalent to the tautology *1 = 1* or *Green is green*. This differs from the factual statements we make about the real world—statements such as "Pine needles are green" or "Chlorophyll molecules reflect green light." These factual statements are approximations. They are technically vague or fuzzy. And they often come juxtaposed with probabilistic uncertainty: "Pine needles are green with high probability." Note that this last statement involves triple uncertainty. There is first the vagueness of green pine needles because there is no bright line between greenness and non-greenness—it is a matter of degree. There is second only a probability whether pine needles have the vague property of greenness. And there is last the magnitude of the probability itself. The magnitude is the vague or fuzzy descriptor "high," because here, too, there is no bright line between high probability and not-high probability.

No one has ever produced a statement of fact that has the same 100 percent binary truth status as a mathematical theorem. Even the most accurate energy predictions of quantum mechanics hold only out to a few decimal places. Binary truth would require getting it right out to infinitely many decimal places.

Most scientists know this and rightly sweat it. The logical premises of a math model only approximately match the world the model purports to model. It is not at all clear how such grounding mismatches propagate through to the model's predictions. Each infected inferential step tends to degrade the confidence of the conclusion, as if multiplying fractions less than one. Modern statistics can appeal to confidence bounds if there are enough samples and if the samples sufficiently approximate the binary assumptions of the model. That at least makes us pay in the coin of data for an increase in certainty.

It is a big step down from such imperfect scientific inference to the approximate syllogistic reasoning of the law. There the disputant insists that similar premises must lead to similar conclusions.

But this similarity involves its own approximate pattern matching of inherently vague patterns of causal conduct or hidden mental states such as intent or foreseeability. The judge's final ruling of "granted" or "denied" resolves the issue in practice. But it is technically a non sequitur. The product of any numbers between zero and one is again always less than one. So the confidence of the conclusion can only fall as the steps in the deductive chain increase. The rap of the gavel is no substitute for proof.

Such approximate reasoning may be as close as we can come to a QED moment when using natural language. The everyday arguments that buzz in our brains hit far humbler logical highs. That is precisely why we all need to prove something at least once—to experience at least one true QED moment. Those rare but godlike tastes of ideal certainty can help keep us from mistaking it for anything else.

OBJECTS OF UNDERSTANDING AND COMMUNICATION

RICHARD SAUL WURMAN

Architect; cartographer; founder, TED Conference; author, 33: Understanding Change & the Change in Understanding

The waking dream I have for my toolkit is one filled with objects of understanding and communication.

The tools in my toolkit respond to me. They nod when I talk, give me evidence of me, and suggest secondary and tertiary journeys that extend my curiosities.

This toolkit is woven of threads of ignorance and stitches of questions that invite knowledge in.

In this weave are maps and patterns with enough stitches to allow me to make the choice, as I wish, to add a tiny drop of superglue.

I want an iPhone/iPad/iMac that nods.

The first movies archived stage shows. The iPad and Kindle archive magazines, newspapers, and books.

I want a new modality with which I can converse at differing levels of complexity, in different languages, and which understands the nuance of my questions.

I want help flying through my waking dreams connecting the threads of these epiphanies.

I believe we are at this cusp.

A first toe in the warm bath of this new modality.

LIFE AS A SIDE EFFECT

CARL ZIMMER
Journalist; author, The Tangled Bank: An Introduction to Evolution*; blogger,* The Loom

It's been more than 150 years since Charles Darwin published the *Origin of Species*, but we still have trouble appreciating the simple, brilliant insight at its core. That is, life's diversity does not exist because it is necessary for living things. Birds did not get wings so that they could fly. We do not have eyes so that we can read. Instead, eyes, wings, and the rest of life's wonder have come about as a side effect of life itself. Living things struggle to survive, they reproduce, and they don't do a perfect job of replicating themselves. Evolution spins off that loop, like heat coming off an engine. We're so used to seeing agents behind everything that we struggle to recognize life as a side effect. I think everyone would do well to overcome that urge to see agents where there are none. It would even help us to understand why we are so eager to see agents in the first place.

THE VEECK EFFECT

GREGORY COCHRAN

Adjunct professor of anthropology, University of Utah; coauthor (with Henry Harpending), The 10,000-Year Explosion: How Civilization Accelerated Human Evolution

There's an invidious rhetorical strategy we've all seen, and I'm afraid that most of us have inflicted it on others as well. I call it the Veeck effect (of the first kind*)—it occurs whenever someone adjusts the standards of evidence in order to favor a preferred outcome.

Why Veeck? Bill Veeck was a flamboyant baseball owner and promoter. In his autobiography, *Veeck—As in Wreck*, he described installing a flexible fence in the right field of the Milwaukee Brewers. At first he put the fence up only when facing a team full of power hitters, but eventually he took it to the limit, putting the fence up when the visitors were at bat and taking it down when his team was.

The history of science is littered with flexible fences. The phlogiston theory predicted that phlogiston would be released when magnesium burned. It looked bad for that theory when experiments showed that burning magnesium became heavier—but its supporters happily explained that phlogiston had negative weight.

Consider Johannes Kepler. He came up with the idea that the distances of the six (known) planets could be explained by nesting the five Platonic solids. It almost worked for Earth, Mars, and Venus but clearly failed for Jupiter. He dismissed the trouble with

---

* If you're wondering about the second Veeck effect, it's the intellectual equivalent of putting a midget up to bat. And that's another essay.

Jupiter, saying, "Nobody will wonder at it, considering the great distance." The theory certainly wouldn't have worked with any extra planets, but fortunately for Kepler's peace of mind, Uranus was discovered well after his death.

The Veeckian urge is strong in every field, but it truly flourishes in the human and historical sciences, where the definitive experiments that would quash such nonsense are often impossible, impractical, or illegal. Nowhere is this tendency stronger than among cultural anthropologists, who at times seem to have no raison d'être other than refurbishing the reputations of cannibals.

Sometimes this has meant denying a particular case of cannibalism—for example, among the Anasazi of the American Southwest. Evidence there has piled up and up. Archaeologists have found piles of human bones with muscles scraped off, split open for marrow, polished by stirring in pots. They have even found human feces with traces of digested human tissue. But that's not good enough. For one thing, this implication of ancient cannibalism among the Anasazi is offensive to their Pueblo descendants, and that somehow trumps mounds of bloody evidence. You would think that the same principle would cause cultural anthropologists to embrace the face-saving falsehoods of other ethnic groups—didn't the South really secede over the tariff? But that doesn't seem to happen.

Some anthropologists have carried the effort further, denying that any culture was ever cannibalistic. They don't just deny Anasazi archaeology; they deny every kind of evidence, from archaeology to historical accounts, even reports from people alive today. When Álvaro de Mendaña discovered the Solomon Islands, he reported that a friendly chieftain threw a feast and offered him a quarter of a boy. Made up, surely. The conquistadors described the Aztecs as a cannibal kingdom. Can't be right, even if the archaeology supports it. When Papuans in Port Moresby volunteered to have a picnic in the morgue—to attract tourists, of course—they were just showing public spirit.

The Quaternary mass extinction, which wiped out much of the world's megafauna, offers paleontologists a chance to crank up their own fences. The large marsupials, flightless birds, and reptiles of Australia disappeared shortly after humans arrived, about fifty thousand years ago. The large mammals of North and South America disappeared about ten thousand years ago—again, just after humans showed up. Moas disappeared within two centuries after Polynesian colonization in New Zealand, while giant flightless birds and lemurs disappeared from Madagascar shortly after humans arrived. What does this pattern suggest as the cause? Why, climate change, of course. Couldn't be human hunters—that's unpossible!

The Veeck effect is even more common in everyday life than it is in science. It's just that we expect more from scientists. But scientific examples are clear-cut, easy to see, and understanding the strategy helps you avoid succumbing to it.

Whenever some administration official says that absence of evidence is not evidence of absence, whenever a psychiatrist argues that Freudian psychotherapy works for some people, even if proven useless on average, Bill Veeck's spirit goes marching on.

SUPERVENIENCE!

JOSHUA GREENE

Cognitive neuroscientist and philosopher, Harvard University

There's a lot of stuff in the world: trees, cars, galaxies, benzene, the Baths of Caracalla, your pancreas, Ottawa, ennui, Walter Mondale. How does it all fit together? In a word . . . supervenience. (Verb form: to *supervene.*) Supervenience is a shorthand abstraction, native to Anglo-American philosophy, that provides a general framework for thinking about how everything relates to everything else.

The technical definition of supervenience is somewhat awkward. Supervenience is a relationship between two sets of properties. Call them Set *A* and Set *B*. The Set *A* properties supervene on the Set *B* properties if and only if no two things can differ in their *A* properties without also differing in their *B* properties.*

This definition, while admirably precise, makes it hard to see what supervenience is really about, which is the relationships among different levels of reality. Take, for example, a computer screen displaying a picture. At a high level, at the level of images, a screen may depict an image of a dog sitting in a rowboat, curled up next to a life vest. The screen's content can also be described as an arrangement of pixels, a set of locations and corresponding colors. The image *supervenes* on the pixels. This is because a screen's image-level properties (its dogginess, its rowboatness) cannot differ from another screen's image-level

---

* Some have pointed out that "supervenience" may also refer to exceptional levels of convenience, as in "New Chinese take-out right around the corner—*Supervenient!*"

properties unless the two screens also differ in their pixel-level properties.

The pixels and the image are, in a very real sense, the same thing. But—and this is key—their relationship is asymmetrical. The image supervenes on the pixels, but the pixels *do not* supervene on the image. This is because screens can differ in their pixel-level properties without differing in their image-level properties. For example, the same image may be displayed at two different sizes or resolutions. And if you knock out a few pixels, it's still the same image. (Changing a few pixels will not protect you from charges of copyright infringement.) Perhaps the easiest way to think about the asymmetry of supervenience is in terms of what determines what. Determining the pixels completely determines the image, but determining the image does not completely determine the pixels.

The concept of supervenience deserves wider currency because it allows us to think clearly about many things, not just about images and pixels. Supervenience explains, for example, why physics is the most fundamental science and why the things that physicists study are the most fundamental things. To many people, this sounds like a value judgment, but it's not, or need not be. Physics is fundamental because everything in the universe, from your pancreas to Ottawa, supervenes on physical stuff. (Or so "physicalists" like me claim.) If there is a universe physically identical to ours, it would include a pancreas just like yours and an Ottawa just like Canada's.

Supervenience is especially helpful when grappling with three contentious and closely related issues: (1) the relationship between science and the humanities, (2) the relationship between the mind and brain, and (3) the relationship between facts and values.

Humanists sometimes perceive science as imperialistic, as aspiring to take over the humanities, to "reduce" everything to electrons, genes, numbers, neurons—and thus to "explain away" all the things that make life worth living. Such thoughts are accompanied by disdain or fear, depending on how credible such ambitions are taken to be. Scientists, for their part, are indeed sometimes imperious, dismissing humanists and their pursuits as childish and unworthy of respect. Supervenience can help us think about how science and the humanities fit together, why science is sometimes perceived as encroaching on the humanist's territory, and the extent to which such perceptions are and are not valid.

It would seem that humanists and scientists study different things. Humanists are concerned with things like love, revenge, beauty, cruelty, and our evolving conceptions of such matters. Scientists study things like electrons and nucleotides. But sometimes it seems like scientists are getting greedy. Physicists aspire to construct a complete physical theory, often called a "Theory of Everything" (TOE). If humanists and scientists study different things, and if physics covers everything, then what is left for the humanists? (Or, for that matter, nonphysicists?)

There is a sense in which a TOE really is a TOE, and there is a sense in which it's not. A TOE is a complete theory of everything upon which everything else *supervenes*. If two worlds are physically identical, then they are also humanistically identical, containing exactly the same love, revenge, beauty, cruelty, and conceptions thereof. But that does not mean that a TOE puts all other theorizing out of business—not by a long shot. A TOE won't tell you anything interesting about *Macbeth* or the Boxer Rebellion.

Perhaps the threat from physics was never all that serious. Today, the real threat, if there is one, is from the behavioral sciences, especially the sciences that connect the kind of "hard" science we all studied in high school to humanistic concerns. In my opinion, three sciences stand out in this regard: behavioral genetics, evolutionary psychology, and cognitive neuroscience. I study moral judgment, a classically humanistic topic. I do this in part by scanning people's brains while they make moral judgments. More recently I've started looking at genes, and my work is guided by evolutionary thinking. My work assumes that the mind supervenes on the brain, and I attempt to explain human values—for example, the tension between individual rights and the greater good—in terms of competing neural systems.

I can tell you from personal experience that this kind of work makes some humanists uncomfortable. During the discussion following a talk I gave at Harvard's Humanities Center, a prominent professor declared that my talk—not any particular conclusion I'd drawn, but the whole approach—made him physically ill. The subject matter of the humanities has always supervened on the subject matter of the physical sciences, but in the past a humanist could comfortably ignore the subvening physical details, much as an admirer of a picture can ignore the pixel-level details. Is that still true? Perhaps it is. Perhaps it depends on one's interests. In any case, it's nothing to be worried sick about.

THE CULTURE CYCLE

HAZEL ROSE MARKUS AND ALANA CONNER
Hazel Rose Markus is the Davis-Brack Professor in the Behavioral Sciences at Stanford University and coauthor (with Paula M. L. Moya), Doing Race: 21 Essays for the 21st Century. *Alana Conner is a science writer, social psychologist, and curator, The Tech Museum, San Jose, California.*

Pundits invoke culture to explain all manner of tragedies and triumphs: why a disturbed young man opens fire on a politician; why African-American children struggle in school; why the United States can't establish democracy in Iraq; why Asian factories build better cars. A quick click through a single morning's media yields the following catch: gun culture, Twitter culture, ethical culture, Arizona culture, always-on culture, winner-take-all culture, culture of violence, culture of fear, culture of sustainability, culture of corporate greed.

Yet no one explains what, exactly, culture is, how it works, or how to change it for the better.

A cognitive tool that fills this gap is the *culture cycle*, a tool that not only describes how culture works but also prescribes how to make lasting change. The culture cycle is the iterative recursive process whereby people create cultures to which they later adapt, and cultures shape people so that they act in ways that perpetuate the cultures.

In other words, cultures and people (and some other primates) make each other up. This process involves four nested levels: individual selves (one's thoughts, feelings, and actions); the everyday practices and artifacts that reflect and shape those selves; the institutions (education, law, media) that afford or discourage those everyday practices and artifacts; and pervasive ideas about what is good, right, and human that both influence and are influenced by all four levels. (See figure below.)

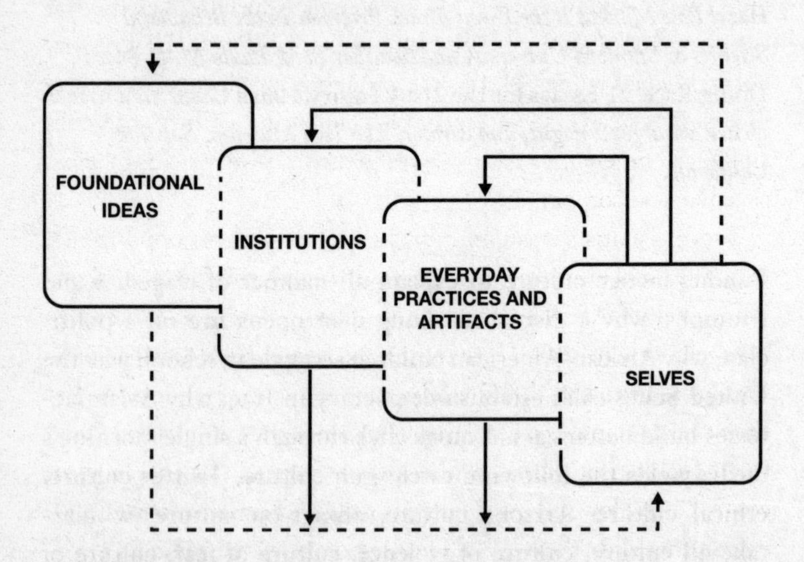

The culture cycle rolls for all types of social distinctions, from the macro (nation, race, ethnicity, region, religion, gender, social class, generation, etc.) to the micro (occupation, organization, neighborhood, hobby, genre preference, family, etc.).

One consequence of the culture cycle is that no action is caused by *either* individual psychological features *or* external influences. Both are always at work. Just as there is no such thing as a culture without agents, there are no agents without culture. Humans are culturally shaped shapers. And so, for example, in the case of a school shooting, it is overly simplistic to ask whether the perpetrator acted because of mental illness, or because of a hostile and bullying school climate, or because he had easy access to a particularly deadly cultural artifact (i.e., a gun), or because institutions encourage that climate and allow access to that artifact, or because pervasive ideas and images glorify resistance and violence. The better question and the one that the culture cycle requires is, How do these four levels of forces interact? Indeed, researchers at the vanguard of public health contend that neither social stressors nor individual vulnerabilities are enough to produce most mental illnesses. Instead, the interplay of biology and culture, of genes and environments, of nature and nurture is responsible for most psychiatric disorders.

Social scientists succumb to another form of this oppositional thinking. For example, in the face of Hurricane Katrina, thousands of poor African-American residents "chose" not to evacuate the Gulf Coast, to quote most news accounts. More charitable social scientists had their explanations ready and struggled to get their variables into the limelight. "Of course they didn't leave," said the psychologists, "because poor people have an external locus of control." Or "low intrinsic motivation." Or "low self-efficacy." "Of course they didn't leave," said the sociologists and political scientists—because their cumulative lack of access to adequate income, banking, education, transportation, health care, police protection, and basic civil rights made staying put their only option. "Of course they didn't leave," said the anthropologists—because their kin networks, religious faith, or historical ties held them there. "Of course they didn't leave," said the economists—because they didn't have the material resources, knowledge, or financial incentives to get out.

The irony in the interdisciplinary bickering is that everyone is mostly right. But they are right in the same way that the blind men touching the elephant in the Indian fable are right: the failure to integrate each field's contributions makes everyone wrong and, worse, not very useful.

The culture cycle illustrates the relationships of these different levels of analyses to one another. Granted, our four-level-process explanation is not as zippy as the single-variable accounts that currently dominate most public discourse. But it's far simpler and accurate than the standard "It's complicated" and "It depends" that more thoughtful experts supply.

Moreover, built into the culture cycle are the instructions for how to reverse-engineer it: A sustainable change at one level usually requires change at all four levels. There are no silver bullets. The ongoing U.S. civil rights movement, for example, requires the opening of individual hearts and minds; the mixing of people as equals in daily life, along with media representations thereof; the reform of laws and policies; and a fundamental revision of our nation's idea of what a good human being is.

Just because people can change their cultures, however, does not mean that they can do so easily. A major obstacle is that most people don't even realize they have cultures. Instead, they think of themselves as standard-issue humans—they're normal; it's all those *other* people who are deviating from the natural, obvious, and *right* way to be.

Yet we are all part of multiple culture cycles. And we should be proud of that fact, for the culture cycle is our smart human trick. Because of it, we don't have to wait for mutation or natural selection to allow us to range farther over the face of the Earth, to extract nutrition from a new food source, to cope with a change in climate. As modern life becomes more complex and social and environmental problems become more widespread and entrenched, people will need to understand the culture cycle and use it skillfully.

PHASE TRANSITIONS AND SCALE TRANSITIONS

VICTORIA STODDEN
Computational legal scholar; assistant professor of statistics, Columbia University

Physicists created the term "phase transition" to describe a change of state in a physical system, such as liquid to gas. The concept has since been applied in a variety of academic circles to describe other types of transformation, from social (think hunter-gatherer to farmer) to statistical (think abrupt changes in algorithm performance as parameters change) but has not yet emerged as part of the common lexicon.

One interesting aspect of the phase transition is that it describes a shift to a state seemingly unrelated to the previous one and hence provides a model for phenomena that challenge our intuition. With knowledge of water only as a liquid, who would have imagined a conversion to gas with the application of heat? The mathematical definition of a phase transition in the physical context is well defined, but even without this precision, this idea can be usefully extrapolated to describe a much broader class of phenomena, particularly those that change abruptly and unexpectedly with an increase in scale.

Imagine points in two dimensions—a spray of dots on a sheet of paper. Now imagine a point cloud in three dimensions—say, dots hovering in the interior of a cube. Even if we could imagine points in four dimensions, would we have guessed that all these points lie on the convex hull of this point cloud? In dimensions greater than three, they always do. There hasn't been a phase transition in the mathematical sense, but as dimension is scaled up, the system shifts in a way we don't intuitively expect.

I call these types of changes "scale transitions," unexpected outcomes resulting from increases in scale. For example, increases in the number of people interacting in a system can produce unforeseen outcomes: The operation of markets at large scales is often counterintuitive. Think of the restrictive effect that rent-control laws can have on the supply of affordable rental housing, or how minimum-wage laws can reduce the availability of low-wage jobs. (James Flynn gives "markets" as an example of a "shorthand abstraction"; here I am interested in the often counterintuitive operation of a market system at large scale.) Think of the serendipitous effects of enhanced communication—for example, collaboration and interpersonal connection generating unexpected new ideas and innovation; or the counterintuitive effect of massive computation in science reducing experimental reproducibility as data and code have proved harder to share than their descriptions. The concept of the scale transition is purposefully loose, designed as a framework for understanding when our natural intuition leads us astray in large-scale situations.

This contrasts with the sociologist Robert K. Merton's concept of "unanticipated consequences," in that a scale transition both refers to a system rather than individual purposeful behavior and is directly tied to the notion of changes due to scale increases. Our intuition regularly seems to break down with scale, and we need a way of conceptualizing the resulting counterintuitive shifts in the world around us. Perhaps the most salient feature of the digital age is its facilitation of huge increases in scale—in data storage, processing power, and connectivity—thus permitting us to address an unparalleled number of problems on an unparalleled scale. As technology becomes increasingly pervasive, I believe scale transitions will become commonplace.

REPLICABILITY

BRIAN KNUTSON
Associate professor of psychology and neuroscience, Stanford University

Since different visiting teachers had promoted contradictory philosophies, the villagers asked the Buddha whom they should believe. The Buddha advised: "When you know for yourselves . . . these things, when performed and undertaken, conduce to well-being and happiness—then live and act accordingly." Such empirical advice might sound surprising coming from a religious leader, but not from a scientist.

"See for yourself" is an unspoken credo of science. It is not enough to run an experiment and report the findings. Others who repeat that experiment must find the same thing. Repeatable experiments are called "replicable." Although scientists implicitly respect replicability, they do not typically explicitly reward it.

To some extent, ignoring replicability comes naturally. Human nervous systems are designed to respond to rapid changes, ranging from subtle visual flickers to pounding rushes of ecstasy. Fixating on fast change makes adaptive sense—why spend limited energy on opportunities or threats that have already passed? But in the face of slowly growing problems, fixation on change can prove disastrous (think of lobsters in the cooking pot or people under greenhouse gases).

Cultures can also promote fixation on change. In science, some high-profile journals, and even entire fields, emphasize novelty, consigning replications to the dustbin of the unremarkable and unpublishable. More formally, scientists are often judged based on their work's novelty rather than its replicability. The increasingly popular "*h*-index" quantifies impact by assigning a number (*h*) which indicates that an investigator has published *h* papers that have been cited *h* or more times (so, Joe Blow has an *h*-index of 5 if he has published five papers, each of which others have cited five or more times). While impact factors correlate with eminence in some fields (e.g., physics), problems can arise. For instance, Dr. Blow might boost his impact factor by publishing controversial (thus, cited) but unreplicable findings.

Why not construct a replicability (or "*r*") index to complement impact factors? As with *h*, *r* could indicate that a scientist has originally documented *r* separate effects that independently replicate *r* or more times (so, Susie Sharp has an *r*-index of 5 if she has published five independent effects, each of which others have replicated five or more times). Replication indices would necessarily be lower than citation indices, since effects have to first be published before they can be replicated, but they might provide distinct information about research quality. As with citation indices, replication indices might even apply to journals and fields, providing a measure that can combat biases against publishing and publicizing replications.

A replicability index might prove even more useful to nonscientists. Most investigators who have spent significant time in the salt mines of the laboratory already intuit that most ideas don't pan out, and those that do sometimes result from chance or charitable interpretations. Conversely, they also recognize that replicability means they're really onto something. Not so for the general public, who instead encounter scientific advances one cataclysmic media-filtered study at a time. As a result, laypeople and journalists are repeatedly surprised to find the latest counterintuitive finding overturned by new results. Measures of replicability could help channel attention toward cumulative contributions. Along those lines, it is interesting to consider applying replicability criteria to public-policy interventions designed to improve health, enhance education, or curb violence. Individuals might even benefit from using replicability criteria to optimize their personal habits (e.g., more effectively dieting, exercising, working, etc.).

Replication should be celebrated rather than denigrated. Often taken for granted, replicability may be the exception rather than the rule. As running water resolves rock from mud, so can replicability highlight the most reliable findings, investigators, journals, and even fields. More broadly, replicability may provide an indispensable tool for evaluating both personal and public policies. As suggested in the *Kalama Sutta*, replicability might even help us decide whom to believe.

AMBIENT MEMORY AND THE MYTH OF NEUTRAL OBSERVATION

XENI JARDIN
Tech culture journalist; partner, contributor, coeditor, Boing Boing; *executive producer, host,* Boing Boing Video

Like others whose early life experiences were punctuated with trauma, my memory has holes. Some of those holes are as wide as years. Others are just big enough to swallow painful incidents that lasted moments but reverberated for decades.

The brain-record of those experiences sometimes submerges, then resurfaces, sometimes submerging again over time. As I grow older, stronger, and more capable of contending with memory, I become more aware of how different my own internal record may be from that of others who lived the identical moment.

Each of us commits our experiences to memory and permanence differently. Time and human experience are not linear, nor is there one and only one neutral record of each lived moment. Human beings are impossibly complex tarballs of muscle, blood, bone, breath, and electrical pulses that travel through nerves and neurons; we are bundles of electrical pulses carrying payloads, pings hitting servers. And our identities are inextricably connected to our environments: No story can be told without a setting.

My generation is the last generation of human beings who were born into a pre-Internet world but who matured in tandem with that great networked hive-mind. In the course of my work online, committing new memories to network mind each day, I have come to understand that our shared memory of events, truths, biography, and fact—all of this shifts and ebbs and flows, just as our most personal memories do.

Ever-edited Wikipedia replaces paper encyclopedias. The chatter of Twitter eclipses fixed-form and hierarchical communication. The news flow we remember from our childhoods, a single voice of authority on one of three channels, is replaced by something hyperevolving, chaotic, and less easily defined. Even the formal histories of a nation may be rewritten by the likes of Wikileaks and its yet unlaunched children.

Facts are more fluid than in the days of our grandfathers. In our networked mind, the very act of observation—reporting or tweeting or amplifying some piece of experience—changes the story. The trajectory of information, the velocity of this knowledge on the network, changes the very nature of what is remembered, who remembers it, and for how long it remains part of our shared archive. There are no fixed states.

So must our notion of memory and record evolve.

The history we are creating now is alive. Let us find new ways of recording memory, new ways of telling the story, that reflect life. Let us embrace this infinite complexity as we commit new history to record.

Let us redefine what it means to remember.

A STATISTICALLY SIGNIFICANT DIFFERENCE IN UNDERSTANDING THE SCIENTIFIC PROCESS

DIANE F. HALPERN

Trustee Professor of Psychology and Roberts Fellow, Claremont McKenna College

Statistically significant difference—it's a simple phrase that is essential to science and has become common parlance among educated adults. These three words convey a basic understanding of the scientific process, random events, and the laws of probability. The term appears almost everywhere that research is discussed—in newspaper articles, advertisements for "miracle" diets, research publications, and student laboratory reports, to name just a few of the many diverse contexts. It is a shorthand abstraction for a sequence of events that includes an experiment (or other research design), the specification of a null and alternative hypothesis, (numerical) data collection, statistical analysis, and the probability of an unlikely outcome. That's a lot of science conveyed in a few words.

It would be difficult to understand the outcome of any research without at least a rudimentary understanding of what is meant by the conclusion that the researchers found or did not find evidence of a "statistically significant difference." Unfortunately, the old saying that "a little knowledge is a dangerous thing" applies to the partial understanding of this term. One problem is that "significant" has a different meaning when used in everyday speech than when used to report research findings.

Most of the time, the word means that something important happened. For example, if a physician told you that you would feel significantly better following surgery, you would correctly infer that your pain would be reduced by a meaningful amount—you would feel less pain. But, when used in "statistically significant difference," "significant" means that the results are unlikely to be due to chance (if the null hypothesis were true); the results themselves may or may not be important. Moreover, sometimes the conclusion will be wrong, because the researchers can assert their conclusion only at some level of probability. "Statistically significant difference" is a core concept in research and statistics, but, as anyone who was taught undergraduate statistics or research methods can tell you, it is not an intuitive idea.

Although "statistically significant difference" communicates a cluster of ideas essential to the scientific process, many pundits would like to see it removed from our vocabulary, because it is frequently misunderstood. Its use underscores the marriage of science and probability theory, and despite its popularity, or perhaps because of it, some experts have called for a divorce, because the term implies something that it should not, and the public is often misled. In fact, experts are often misled as well. Consider this hypothetical example: In a well-done study that compares the effectiveness of two drugs relative to a placebo, it is possible that Drug X is statistically significantly different from a placebo and Drug Y is not, yet Drugs X and Y might not be statistically significantly different from each other. This could result when Drug X is statistically different from placebo at a probability level of $p < .04$ but Drug Y is statistically significantly different from a placebo only at a probability level of $p < .06$, which is higher than most a priori levels used to test for statistical significance. If reading about this makes your head hurt, you are among the masses who believe they understand this critical shorthand phrase which is at the heart of the scientific method but who actually may have only a shallow level of understanding.

A better understanding of the pitfalls associated with this term would go a long way toward improving our cognitive toolkits. If common knowledge of what this term means included the ideas that (a) the findings may not be important, and (b) conclusions based on finding or failure to find statistically significant differences may be wrong, then we would have substantially advanced our general knowledge. When people read or use the term "statistically significant difference," it is an affirmation of the scientific process, which, for all its limitations and misunderstandings, is a substantial advance over alternative ways of knowing about the world. If we could just add two more key concepts to the meaning of that phrase, we could improve how the general public thinks about science.

THE DECE(I)BO EFFECT

BEATRICE GOLOMB

Associate professor of medicine, University of California–San Diego

The Dece(i)bo effect (think a portmanteau of "deceive" and "placebo") refers to the facile application of constructs—without unpacking the concept and the assumptions on which it relies—in a fashion that, rather than benefiting thinking, leads reasoning astray.

Words and phrases that capture a concept enter common parlance: Ockham's razor, placebo, Hawthorne effect. Such phrases and code words, in principle, facilitate discourse—and can indeed do so. Deploying the word or catchphrase adds efficiency to the interchange by obviating the need for a pesky review of the principles and assumptions encapsulated in the word. Unfortunately, bypassing the need to articulate the conditions and assumptions on which validity of the construct rests may lead to bypassing consideration of whether these conditions and assumptions legitimately apply. Use of the term can then, far from fostering sound discourse, serve to undermine it.

Take, for example, the "placebo" and "placebo effects." Unpacking the terms, a placebo is defined as something physiologically inert but believed by the recipient to be active or possibly so. The term "placebo effect" refers to improvement of a condition when someone has received a placebo—improvement due to the effects of expectation/suggestion.

With these terms ensconced in the vernacular, dece(i)bo effects associated with them are much in evidence. Key presumptions regarding placebos and placebo effects are more typically wrong than not.

1. When hearing the word "placebo," scientists often presume "inert" without stopping to ask, What *is* that allegedly physiologically inert substance? Indeed, in principle, what substance *could* it be? There isn't anything known to be physiologically inert.

   There are no regulations about what constitute placebos, and their composition—commonly determined by the manufacturer of the drug under study—is typically undisclosed. Among the uncommon cases where placebo composition has been noted, there are documented instances in which the placebo composition apparently produced spurious effects. Two studies used corn-oil and olive-oil placebos for cholesterol-lowering drugs. One noted that the "unexpectedly" low rate of heart attacks in the control group may have contributed to failure to see a benefit from the cholesterol drug. Another study noted the "unexpected" benefit of a drug to gastrointestinal symptoms in cancer patients. But cancer patients bear an increased likelihood of lactose intolerance—and the placebo was lactose, a "sugar pill." When the term "placebo" substitutes for actual ingredients, any thinking about how the composition of the control agent may have influenced the study is circumvented.

2. Because there are many settings in which people with a problem, given a placebo, report sizable improvement on average when they are queried (see #3), many scientists have accepted that "placebo effects"—of suggestion—are both substantial and widespread in the scope of what they benefit.

   The Danish researchers Asbjørn Hróbjartsson and Peter C. Götzsche conducted a systematic review of studies that compared a placebo to no treatment. They found that the placebo generally does . . . nothing. In most instances, there is no placebo effect. Mild "placebo effects" are seen, in the short term, for pain and anxiety. Placebo effects for pain are reported to be blocked by naloxone, an opiate antagonist—specifically implicating endogenous opiates in pain placebo effects, which would not be expected to benefit every possible outcome that might be measured.

3. When hearing that people given a placebo report improvement, scientists commonly presume this must be due to the "placebo effect," the effect of expectation/suggestion. However, the effects are usually something else entirely—for instance, the natural history of the disease, or regression to the mean. Consider a distribution such as a bell curve. Whether the outcome of interest is the reduction of pain, blood pressure, cholesterol, or something else, people are classically selected for treatment if they are at one end of the distribution—say, the high end. But these outcomes are quantities that vary (for instance because of physiological variation, natural history, measurement error, etc.), and on average the high values will vary back down—a phenomenon termed "regression to the mean" that operates, placebo or no. (Hence, the Danish researchers' findings.)

A different dece(i)bo problem beset Ted Kaptchuk's recent Harvard study in which researchers gave a placebo, or nothing, to people afflicted with irritable bowel syndrome. They administered the placebo in a bottle boldly labeled "Placebo" and advised patients that they were receiving placebos, which were known to be potent. The thesis was that one might harness the effects of expectation honestly, without deception, by telling subjects how powerful placebos in fact were—and by developing a close relationship with subjects. Researchers met repeatedly with subjects, gained subjects' appreciation for their concern, and repeatedly told subjects that placebos are powerful. Those placed on the placebo obliged the researchers by telling them that they got better, more so than those on nothing. The scientists attributed this to a placebo effect.

But what's to say that the subjects weren't simply telling the scientists what they thought the scientists wished to hear? Denise Grady, writing for the *New York Times*, has noted: "Growing up, I got weekly hay fever shots that I don't think helped me at all. But I kept hoping they would, and the doctor was very kind, so whenever he asked if I was feeling better, I said yes . . ." Such desire to please (a form, perhaps, of "social approval" reporting bias) made for fertile ground in which to operate and create what was interpreted as a placebo effect, which implies actual subjective benefit to symptoms. One wonders whether so great an error of presumption would operate were there not an existing term ("placebo effect") to signify the interpretation the Harvard group chose among the suite of other compelling possibilities.

Another explanation consistent with these results is specific physiological benefit. The Kaptchuk study used a nonabsorbed fiber—microcrystalline cellulose—as the placebo that subjects were told would be effective. The authors are applauded for disclosing its composition. But other nonabsorbed fibers benefit both constipation and diarrhea—symptoms of irritable bowel—and are prescribed for that purpose; psyllium is an example. Thus, specific physiological benefit of the "placebo" to the symptoms cannot be excluded.

Together these points illustrate that the term "placebo" cannot be presumed to imply "inert" (and generally does not); and that when studies see large benefit to symptoms in patients treated with a placebo (a result expected from distribution considerations alone), one cannot infer that these arose from suggestion.

Thus, rather than facilitating sound reasoning, evidence suggests that in many cases, including high-stakes settings in which inferences may propagate to medical practice, substitution of a term—here, "placebo" and "placebo effect"—for the concepts they are intended to convey may actually thwart or bypass critical thinking about key issues, with implications for fundamental concerns of us all.

ANTHROPOPHILIA

ANDREW REVKIN
Journalist; environmentalist; writer, New York Times*'s* Dot Earth *blog; author,* The North Pole Was Here

To sustain progress on a finite planet that is increasingly under human sway but also full of surprises, what is needed is a strong dose of *anthropophilia*. I propose this word as shorthand for a rigorous and dispassionate kind of self-regard, even self-appreciation, to be employed when individuals or communities face consequential decisions attended by substantial uncertainty and polarizing disagreement.

The term is an intentional echo of E. O. Wilson's valuable effort to nurture *biophilia*, the part of humanness that values and cares for the facets of the nonhuman world we call nature. What's been missing too long is an effort to fully consider, even embrace, the human role *within* nature and—perhaps more important still—to consider our own inner nature as well.

Historically, many efforts to propel a durable human approach to advancement were shaped around two organizing ideas: "Woe is me" and "Shame on us," with a good dose of "Shame on you" thrown in.

The problem?

Woe is paralytic, while blame is both divisive and often misses the real target. (Who's the bad guy, BP or those of us who drive and heat with oil?) Discourse framed around those concepts too often produces policy debates that someone once described to me, in the context of climate, as "blah, blah, blah, *bang*." The same phenomenon can as easily be seen in the unheeded warnings leading to the September 11 attacks and to the most recent financial implosion.

More fully considering our nature—both the "divine and felonious" sides, as Bill Bryson has summed us up—could help identify certain kinds of challenges that we *know* we'll tend to get wrong. The simple act of recognizing such tendencies could help refine how our choices are made—at least giving slightly better odds of getting things a little less wrong the next time. At the personal level, I know that when I cruise into the kitchen tonight I'll tend to prefer reaching for a cookie instead of an apple. By preconsidering that trait, I may have a slightly better chance of avoiding a couple hundred unnecessary calories.

Here are a few instances where this concept is relevant on larger scales.

There's a persistent human pattern of not taking broad lessons from localized disasters. When China's Sichuan province was rocked by a severe earthquake, tens of thousands of students (and their teachers) died in collapsed schools. Yet the American state of Oregon, where more than a thousand schools are already known to be similarly vulnerable when the great Cascadia fault off the Northwest Coast next heaves, still lags terribly in speeding investments in retrofitting. Sociologists understand, with quite a bit of empirical backing, why this disconnect exists even though the example was horrifying and the risk in Oregon is about as clear as any scientific assessment can be. But does that knowledge of human biases toward the "near and now" get taken seriously in the realms where policies are shaped and the money to carry them out is authorized? Rarely, it seems.

Social scientists also know, with decent rigor, that the fight over human-driven global warming—both over the science and policy choices—is largely cultural. As in many other disputes (consider health care), the battle is between two fundamental subsets of human communities—communitarians (aka liberals) and individualists (aka libertarians). In such situations, a compelling body of research has emerged showing that information is fairly meaningless. Each group selects information to reinforce a position, and there are scant instances where information ends up shifting a position. That's why no one should expect the next review of climate science from the Intergovernmental Panel on Climate Change to suddenly create a harmonious path forward.

The more such realities are recognized, the more likely it is that innovative approaches to negotiation can build from the middle, instead of arguing endlessly from the edge. The same body of research on climate attitudes, for example, shows far less disagreement on the need for advancing the world's limited menu of affordable energy choices.

The physicist Murray Gell-Mann has spoken often of the need, when faced with multidimensional problems, to take a "crude look at the whole"—a process he has even given an acronym, CLAW. It's imperative, where possible, for that look to include an honest analysis of the species doing the looking as well.

There will never be a way to invent a replacement for, say, the United Nations or the House of Representatives. But there is a ripe opportunity to try new approaches to constructive discourse and problem solving, with the first step being an acceptance of our humanness, for better and worse.

That's anthropophilia.

A SOLUTION FOR COLLAPSED THINKING: SIGNAL DETECTION THEORY

MAHZARIN R. BANAJI

Richard Clarke Cabot Professor of Social Ethics, Department of Psychology, Harvard University

We perceive the world through our senses. The brain-mediated data we receive in this way form the basis of our understanding of the world. From this becomes possible the ordinary and exceptional mental activities of attending, perceiving, remembering, feeling, and reasoning. Via these mental processes, we understand and act on the material and social world.

In the town of Pondicherry in South India, where I sit as I write this, many do not share this assessment. There are those, including some close to me, who believe there are extrasensory paths to knowing the world that transcend the five senses, that untested "natural" foods and methods of acquiring information are superior to those based in evidence. On this trip, for example, I learned that they believe that a man has been able to stay alive without any caloric intake for months (although his weight falls, but only when he is under scientific observation).

Pondicherry is an Indian Union Territory that was controlled by the French for three hundred years (staving off the British in many a battle right outside my window) and that the French held on to until a few years after Indian independence. It has, in addition to numerous other points of attraction, become a center for those who yearn for spiritual experience, attracting many (both whites and natives) to give up their worldly lives to pursue the advancement of the spirit, undertake bodily healing, and invest in good works on behalf of a larger community.

Yesterday I met a brilliant young man who had worked as a lawyer for eight years and now lives in an ashram and works in its book-sales division. "Sure," you retort, "the legal profession would turn any good person toward spirituality," but I assure you that the folks here have given up wealth and a wide variety of professions to pursue this manner of life. The point is that seemingly intelligent people seem to crave nonrational modes of thinking.

I do not mean to pick on any one city, and certainly not this unusual one in which so much good effort is spent on the arts and culture and social upliftment of the sort we would admire. But this is also a city that attracts a particular type of European, American, and Indian—those whose minds seem more naturally prepared to believe that herbs do cure cancer and standard medical care is to be avoided (until one desperately needs chemotherapy), that Tuesdays are inauspicious for starting new projects, that particular points in the big toe control the digestive system, that the position of the stars at the time of their birth led them to Pondicherry through an inexplicable process emanating from a higher authority and through a vision from "the Mother," a deceased Frenchwoman who dominates the ashram and surrounding area in death more thoroughly than many skilled politicians do during their terms in office.

These types of beliefs may seem extreme, but they are not considered so in most of the world. Change the content, and the underlying false manner of thinking is readily observed just about anywhere. The twenty-two inches of new snow that fell recently where I live in the United States will no doubt bring forth beliefs of a god angered by crazy scientists touting global warming.

As I contemplate the single most powerful tool that could be put into our toolkits, it is the simple and powerful concept of "signal detection." In fact, the *Edge* Question this year happens to be one I've contemplated for a while. I use David Green and John Swets's *Signal Detection Theory and Psychophysics* as the prototype, although the idea has its origins in earlier work among scientists concerned with the influence of photon fluctuations on visual detection and of sound waves on audition.

The idea underlying the power of signal-detection theory is simple: The world provides us with noisy, not pure, data. Auditory data, for instance, are degraded for a variety of reasons having to do with the physical properties of the communication of sound. The observing organism has properties that further affect how those data will be experienced and interpreted, such as auditory acuity; the circumstances under which the information is being processed (e.g., during a thunderstorm); and motivation (e.g., disinterest). Signal-detection theory allows us to put both aspects of the stimulus and the respondent together to understand the quality of the decision that will result, given the uncertain conditions under which data are transmitted both physically and psychologically.

To understand the crux of signal-detection theory, each event of any data impinging on the receiver (human or other) is coded into four categories, providing a language to describe the decision. One dimension concerns whether an event occurred or not (was a light flashed or not?); the other dimension concerns whether the human receiver detected it or not (was the light seen or not?). This gives us a 2 x 2 table of the sort laid out below, but it can be used to configure many different types of decisions. For example, were homeopathic pills taken or not? Did the disease get cured or not?

Did the event occur?		
	Yes	No
Yes	Hit	False Alarm
Was the event detected?		
No	Miss	Correct Rejection

Hit: A signal is present, and the signal is detected (correct response)
False Alarm: No signal is presented, but a signal is detected (incorrect response)
Miss: A signal is present, but no signal is detected (incorrect response)
Correct Rejection: No signal is presented, and no signal is detected (correct response)

If the signal is clear, like a bright light against a dark background, and the decision maker has good visual acuity and is motivated to watch for the signal, we should see a large number of Hits and Correct Rejections and very few False Alarms and Misses. As these properties change, so does the quality of the decision. It is under ordinary conditions of uncertainty that signal-detection theory yields a powerful way to assess the stimulus and respondent qualities, including the respondent's idiosyncratic criterion (or cutting score) for decision making.

Signal-detection theory has been applied in areas as diverse as locating objects by sonar, the quality of remembering, the comprehension of language, visual perception, consumer marketing, jury decisions, price predictions in financial markets, and medical diagnoses. The reason signal-detection theory should be in the toolkit of every scientist is because it provides a mathematically rigorous framework for understanding the nature of decision processes. The reason its logic should be in the toolkit of every thinking person is because it forces a completion of the four cells when analyzing the quality of any statement, such as "Good management positions await Sagittarius this week."

EVERYDAY APOPHENIA

DAVID PIZARRO
Assistant professor, Department of Psychology, Cornell University

The human brain is an amazing pattern-detecting machine. We possess a variety of mechanisms that allow us to uncover hidden relationships between objects, events, and people. Without these, the sea of data hitting our senses would surely appear random and chaotic. But when our pattern-detection systems misfire, they tend to err in the direction of perceiving patterns where none actually exist.

The German neurologist Klaus Conrad coined the term "apophenia" to describe this tendency in patients suffering from certain forms of mental illness. But it is increasingly clear from a variety of findings in the behavioral sciences that this tendency is not limited to ill or uneducated minds; healthy, intelligent people make similar errors on a regular basis. A superstitious athlete sees a connection between victory and a pair of socks; a parent refuses to vaccinate her child because of a perceived causal connection between inoculation and disease; a scientist sees hypothesis-confirming results in random noise; and thousands of people believe the random "shuffle" function on their music software is broken because they mistake spurious coincidence for meaningful connection.

In short, the pattern-detection responsible for so much of our species' success can just as easily betray us. This tendency to oversee patterns is likely an inevitable by-product of our adaptive pattern-detecting mechanisms. But the ability to acknowledge, track, and guard against this potentially dangerous tendency would be aided if the simple concept of "everyday apophenia" were an easily accessible concept.

A COGNITIVE TOOLKIT FULL OF GARBAGE

ERNST PÖPPEL

Neuroscientist; chairman of the Human Science Center, Munich University; author, Mindworks: Time and Conscious Experience

To get rid of garbage is essential. Also mental garbage. Cognitive toolkits are filled with such garbage simply because we are victims of ourselves. We should regularly empty this garbage can or, if we enjoy sitting in garbage, we'd better check how "shorthand abstractions" (SHAs) limit our creativity (certainly itself a SHA). Why is the cognitive toolkit filled with garbage?

Let us look back in history (SHA): Modern science (SHA) can be said to have started in 1620 with *Novum Organum* ("New Instrument"), by Francis Bacon. It should impress us today that his analysis (SHA) begins with a description (SHA) of four mistakes we run into when we do science. Unfortunately, we usually forget these warnings. Francis Bacon argued that we are, first, victims of evolution (SHA)—that is, that our genes (SHA) define constraints that necessarily limit insight (SHA). Second, we suffer from the constraints of imprinting (SHA); the culture (SHA) we live in provides a frame for epigenetic programs (SHA) that ultimately define the structure (SHA) of neuronal processing (SHA). Third, we are corrupted by language (SHA), because thoughts (SHA) cannot be easily transformed into verbal expressions. Fourth, we are guided, or even controlled, by theories (SHA), be they explicit or implicit.

What are the implications for a cognitive toolkit? We are caught, for instance, in a language trap. On the basis of our evolutionary heritage, we have the power of abstraction (SHA), but this has, in spite of some advantages we brag about (to make us seem superior to other creatures), a disastrous consequence: Abstractions are usually represented in words; apparently we cannot do otherwise. We have to "ontologize"; we invent nouns to extract knowledge (SHA) from processes (SHA). (I do not refer to the powerful pictorial shorthand abstractions.) Abstraction is obviously complexity reduction (SHA). We make it simple. Why do we do this? Evolutionary heritage dictates rapidity. However, speed may be an advantage for a survival toolkit but not for a cognitive toolkit. It is a categorical error (SHA) to confuse speed in action with speed in thinking. The selection pressure for speed invites us to neglect the richness of facts. This pressure allows the invention (SHA) of a simple, clear, easy-to-understand, easy-to-refer-to, easy-to-communicate shorthand abstraction. Thus, because we are a victim of our biological past, and as a consequence a victim of ourselves, we end up with shabby SHAs, having left behind reality. If there is one disease all humans share, it is "monocausalitis," the motivation (SHA) to explain everything on the basis of just one cause. This may be a nice intellectual exercise, but it is simply misleading.

Of course we depend on communication (SHA), and this requires verbal references usually tagged with language. But if we do not understand, within the communicative frame or reference system (SHA), that we are a victim of ourselves by "ontologizing" and continually creating "practical" SHAs, we simply use a cognitive toolkit of mental garbage.

Is there a pragmatic way out, other than to radically get rid of mental garbage? Yes, perhaps: Simply not using the key SHAs explicitly in one's toolkit. Working on consciousness, don't use (at least for one year) the SHA "consciousness." If you work on the "self," never refer explicitly to self. Going through one's own garbage, one discovers many misleading SHAs, like just a few in my focus of attention (SHA): the brain as a net, localization of function, representation, inhibition, threshold, decision, the present. An easy way out is, of course, to refer to some of these SHAs as metaphors (SHA), but this, again, is evading the problem (SHA). I am aware of the fact (SHA) that I am also a victim of evolution, and to suggest "garbage" as a SHA also suffers from the same problem; even the concept of garbage required a discovery (SHA). But we cannot do otherwise than simply be aware of this challenge (SHA), that the content of the cognitive toolkit is characterized by self-referentiality (SHA)—that is, by the fact that the SHAs define themselves by their unreflected use.

ACKNOWLEDGMENTS

Thanks to Steven Pinker for suggesting this year's *Edge* Question and to Daniel Kahneman for advice on its presentation. Thanks to Peter Hubbard of HarperCollins for his continued support. And thanks to Sara Lippincott for her thoughtful and meticulous editing.

INDEX